The Practical Einstein

The Practical Einstein

Experiments, Patents, Inventions

József Illy

The Johns Hopkins University Press
Baltimore

© 2012 The Johns Hopkins University Press
All rights reserved. Published 2012
Printed in the United States of America on acid-free paper

Johns Hopkins Paperback edition, 2013
9 8 7 6 5 4 3 2 1

The Johns Hopkins University Press
2715 North Charles Street
Baltimore, Maryland 21218-4363
www.press.jhu.edu

The Library of Congress has cataloged the hardcover edition
of this book as follows:

Illy, József, 1933–
   The practical Einstein : experiments, patents, inventions / József Illy.
       p. cm.
   Includes bibliographical references and index.
   ISBN-13: 978-1-4214-0457-8 (hdbk. : acid-free paper)
   ISBN-10: 1-4214-0457-5 (hdbk. : acid-free paper)
   ISBN-13: 978-1-4214-0533-9 (electronic)
   ISBN-10: 1-4214-0533-4 (electronic)
   1. Einstein, Albert, 1879–1955—Influence.   2. Inventions.   3. Physics—
Experiments.   I. Title.
   QC16.E5I45 2012
   530.078—dc23   2011034750

A catalog record for this book is available from the British Library.

ISBN-13: 978-1-4214-1171-2
ISBN-10: 1-4214-1171-7

Special discounts are available for bulk purchases of this book. For more
information, please contact Special Sales at 410-516-6936 or specialsales@
press.jhu.edu.

The Johns Hopkins University Press uses environmentally friendly book
materials, including recycled text paper that is composed of at least 30
percent post-consumer waste, whenever possible.

To Marci with I+Sz

Contents

Preface

The scope of the experiments, opinions for patent cases, and inventions in which Albert Einstein participated shows how multifarious his activities were, how deeply he was involved in searching for various technological solutions, and how wide his knowledge of physics was outside the fields for which he has been so famous.

I am not a pioneer in exploring this topic; the first book dedicated to Einstein's practical ideas is Frenkel and Yavelov's account.[1] Wolfgang Graff's dissertation concentrates on Einstein's inventions made before Einstein emigrated to the United States. Graff is the ideal historian to discuss Einstein and Leó Szilárd's refrigerators, as he is a specialist in cooling technology.[2] Michael Eckert presents a broad picture of aeronautics around the First World War,[3] and Jobst Broelmann does the same for the development of the gyrocompass, two other fields in which Einstein tested his mettle.[4] Dieter Lohmeier and Bernhardt Schell analyze Einstein's contribution to the gyro compass by publishing and annotating his correspondence with Hermann Anschütz-Kaempfe and colleagues.[5] Finally, Matthew Trainer published a concise report on Einstein's selected inventions.[6]

My primary source of information in researching the book was the Albert Einstein Archives at the Hebrew University in Jerusalem. This archive, as with archives in general, can be searched for names of correspondents and dates, but rarely for the content of the documents. This entails no problem when the persons who collaborated with Einstein in technical matters are widely known, such as Leó Szilárd, Gustav Bucky, or Rudolf Goldschmidt. When, however, Einstein proposes an experiment or is approached for an opinion, sometimes in letters exchanged with friends or family members, only a systematic reading of the documents will reveal the contents. Such reading has already been done for the period 1879–1922, and the results published in the *Collected Papers of Albert Einstein*. These volumes are the reason for the abundance of information from this early period, not only in the present book but also in the Einstein biographies published in the past decades.

Here and there lie hints at opinions, experiments, and inventive ideas that are not discussed in detail. As examples, Fritz Haber, Einstein's close acquaintance and colleague, encouraged I. Rosenberg, member of the board of the Allgemeine Gesellschaft für Chemische Industrie, to seek Einstein's expert opinion on a revocatory action of Konrad Kubierschky on optical and physicochemical phenomena.[7] Einstein was ready to contribute[8] and prepared his opinion before June 19, 1919, but the opinion is not available. Kubierschky owned patents on distillation towers, absorption cooling, heating, and other inventions in the first decades of the twentieth century, but none of these provides useful information on Einstein's role.

A brief sentence in a letter of Hermann Anschütz-Kaempfe to Einstein notes that Einstein had proposed using two gyroscopes rotating in opposite directions for an artificial horizon for aircraft and ship.[9] Again, no details are known. Another sentence in a letter to Elsa Einstein talks about an experiment, proposed by Einstein and to be performed by Pieter Zeeman.[10] And a comment in another letter envisages an experiment at the University of Kiel, with Walther Kossel.[11]

Time and again, secondary literature asserts that Einstein, together with a painter and a dentist, participated in the development of an apparatus for copying artistic drawings. Whether it is true, and whether the painter was Emil Orlik and the dentist Josef Grünberg, are questions whose answers may still be buried in the abyss of the Einstein Archives. But in the infrequent letters they exchanged, there is no mention of such an invention.

The most serious limitation in writing this book, however, has been that I am not an engineer in any field Einstein was active in. I therefore welcome any critical comments, suggestions, or discoveries of new documents. Please send them to me at illy@einstein.caltech.edu.

English translations of many of the quotations in the book are from the volumes of translations accompanying the documentary edition of the *Collected Papers of Albert Einstein* and also from *Einstein, Anschütz and the Kiel Gyro Compass*[12] and *The Born-Einstein Letters 1916–1955*.[13]

I thank Trevor Lipscombe, former editor in chief of the Johns Hopkins University Press, for proposing the topic and for his erudite stylistic corrections. It was also a pleasure to work with Greg Nicholl, assistant acquisitions editor, and with Michele Callaghan. I deeply appreciate the work of copy editor Brian MacDonald. His meticulous revision was all the more tedious a task because the text was not from the pen of a native speaker.

I express my sincere thanks to Diana Buchwald, director and general editor of the Einstein Papers Project, for encouragement and for permission to use the facilities of the Einstein Papers Project.

Jane Dietrich's critical remarks contributed to a clearer exposition of my intention in presenting the book's themes. She also used her skill and many years of experience to edit the manuscript for grammar and wording. Sincere thanks are due to my colleague Daniel Kennefick for his help in some calculations. My colleagues Osik Moses, Carol Chaplin, Jennifer Nollar, and Lernik Ohanian gave me a hand in obtaining source materials.

For help in finding documents and discussing them I am indebted to Jobst Broelmann, Michael Eckert, Wilhelm Füßl, and Eva A. Mayring, Deutsches Museum, Munich; László Futó, Forestry College Székesfehérvár; Heinz Fütterer, Deutsches Zentrum für Luft- und Raumfahrt, Göttingen; Scott J. Gilbert, American Association of University Women, National Foundation, Alice Ann Leidel Library and Marion Talbot Archives, Washington, D.C.; Zofia Gołąb-Meyer, Uniwersytet Jagielloński, Cracow; David Leidenborg, AB Electrolux, Stockholm; Robert Michaelson, Northwestern University, Evanston, Illinois; Gene Morris and Barry L. Zerry, National Archives and Records Administration, College Park, Maryland; Florian Rohm, European Patent Office, Munich; Bernhardt Schell and Martin Richter, Raytheon Anschütz GmbH, Kiel; Jörg Schmalfuß, Deutsches Technikmuseum, Berlin; Frederic D. Schwarz, *Invention and Technology*; Ronald Smith Jr., Burtonsville, Maryland; John Stachel, Boston University, Boston, Massachusetts, and Jeroen Tromp, California Institute of Technology, Pasadena, California.

I am indebted to Roni Grosz, curator of the Albert Einstein Archives, Jerusalem; Jessika Wichner, Deutsches Zentrum für Luft- und Raumfahrt, Zentrales Archiv, Göttingen; and the George Grantham Bain Collection, Library of Congress, Washington, D.C., for illustrations. I also made ample use of the excellent Web site of the European Patent Office.

I am most grateful to my wife, Marci, who tolerated my seclusion for months on end and compassionately followed my wrestling with the sources.

The Practical Einstein

Introduction

Albert Einstein was born on March 14, 1879, in Ulm, in southern Germany. His uncle and his father had a firm producing electrotechnical apparatus: generators, transformers, measuring devices, and electric lighting networks for towns. They started the enterprise in Germany, then moved to northern Italy, to Pavia (fig. I.1). Einstein's father, Hermann, was the business manager, his uncle, Jacob, the specialist engineer.[1] Originally, Einstein also wanted to be an engineer, and his inclination to technical matters was observed by his uncle: "There is something wonderful about my nephew," he remarked. "Where I and my assistant engineer had cudgeled our brains all day long, the young guy figured the whole thing out in less than a quarter of an hour. He will get on in the world."[2]

After Einstein passed the final examinations at high school, he enrolled in the Zurich Polytechnic (the "Poly") in 1896. He graduated from it in 1900 with a diploma that qualified him to teach high school mathematics and physics. He prepared an experimental thesis in Professor Heinrich Friedrich Weber's laboratory, apparently on heat conduction, and after that planned to study the Thomson effect by use of a novel method for determining the temperature dependence of thermal conductivity.[3] No details are known. Even though

Elektrotechnische Fabrik

J. Einstein & Cie.

München.

Ausführung
elektrischer
Beleuchtungs-
anlagen
in jedem Um-
fange.

Ausführung
elektrischer
Kraftüber-
tragungs-
anlagen
jeder Grösse.

Fabrikation

von

Dynamo-Maschinen

für

Beleuchtung, Kraftübertragung und Elektrolyse,

Bogenlampen, Elektrizitätszählern,

Mess- und Regulirapparaten.

Figure I.1. Advertisement for the company run by Einstein's father and uncle.

these experiments constituted steps toward the solution of a fundamental theoretical problem, the relationship between thermal and electrical phenomena and, by extension, the structure of matter, he was also fascinated by the "direct contact with experience."[4]

After several attempts to get a job as a teacher, he applied for a position at the Swiss Patent Office. In 1902 he was hired as technical expert third class, with the duty, as his director, Friedrich Haller taught him, to think that everything the inventor says is wrong.

Einstein was able to recognize the benefits of the "practical professions":

> The work on the definitive formulation of technical patents was a real blessing for me. It forced extensive thinking and also brought important stimuli to physical thinking. . . . For the academic career puts the young in a kind of constrained position to produce scientific works in impressive amounts—a seduction to superficiality. . . . In addition, most of the practical professions are such that a man with normal abilities is supposed to produce what is expected from him. In his social existence he does not depend on peculiar inspirations. If he has a deeper scientific curiosity, he can immerse himself in his favorite problem besides his compulsory work. The fear need not weigh upon him that his efforts may work to no end.[5]

He was writing here about himself. Besides his work in the patent office, he fathered two sons, completed a Ph.D. dissertation, and wrote reviews of scientific papers, as well as papers of his own on electrodynamics of moving bodies (later called special relativity); on emission and absorption of light, which contained the hypothesis that light consists of energy quanta (later called photons); on the explanation of Brownian motion (a proof that molecules have a real existence); and on the equivalence of energy and mass, which led to the famous formula $E = mc^2$. He also took the first steps toward general relativity during this time.

Einstein was promoted from provisional to permanent status in 1904, though Haller added the caveat that "he has not yet familiarized himself with the study of machine construction."[6] In less than two years, Haller changed his opinion, for Einstein has "increasingly familiarized himself with technical problems, so that he now handles the most difficult technical patent applications with great success and belongs to the most prized examiners in the agency."[7]

He received his doctorate from the University of Zurich in 1906 and became *Privatdozent* at the University of Bern in 1908. After having been appointed extraordinary professor of theoretical physics at the University of Zurich, he left the patent office.

Einstein did not consider entering a career in technology. "The thought of having to expend my creative energy on things that make practical everyday life even more refined, with the goal of bleak capital gain, was intolerable to me," he wrote to his friend Heinrich Zangger in 1918.[8] These words were preceded by remarks about his eighteen-year-old son, Hans Albert: "Albert is already starting to think quite amusingly, oddly enough, about technical questions. But, after all, I am glad for any intellectual activity, even if it clings to petit bourgeois views. Perhaps he will eventually realize how superfluous those many conveniences are!" We will, however, see that the time was approaching when the father Einstein did yield to petit bourgeois views and coquetted with expending his creative energy on things practical for everyday life.

As a full-time scientist, Einstein's achievements brought him spectacular success. By 1909, he was nominated for the Nobel Prize in Physics by a previous recipient. After having spent a short time at the University of Prague and at his alma mater, the Poly (which had grown into the Swiss Federal Institute of Technology), as full professor, the thirty-four-year-old Einstein was invited to join the Prussian Academy of Sciences, one of the most prestigious scientific bodies of the time, with no obligations but research.

In Berlin, he gave final form to his general theory of relativity, and when one of its consequences, the light deflection by the sun, was confirmed by a British group in 1919, his name conquered the front pages of the world's most famous newspapers. He became a world celebrity.

His next goal was to create a theory that would explain not only the large-scale structure of the universe as general relativity did but also how matter is built from atoms and its constituents: a theory unifying general relativity and electromagnetic theory. It took six years after 1916, the year he published the final form of general relativity, before he dared turn to the scientific public with a paper on a kind of a unified theory.[9] Those six hard years had moments of frustration, disappointment, desperation. He felt himself old for creating something comparable to the achievements of his youth.[10] "What do you want from an old wreck or empty eggshell like me?" he wrote to Heinrich Zangger. "The intellect gets crippled and one's strength dwindles away, but glittering renown is still draped around the calcified shell."[11] Next year, when Arnold Sommerfeld requested a lecture at the University of Munich, he apologized for not accepting the invitation by saying, "I don't really have anything new to present, just a rehash."[12] And even in 1922 he felt that "what I could say of [my] science is noisily twittered . . . by the sparrows on the roofs."[13] As his autobiographical notes suggest, he was now constrained to produce scientific articles in large quantities, and so fondly remembered the Swiss Patent Office as "that worldly cloister where I concocted my finest ideas."[14]

Such nostalgia moved him to propose a similar career to a young colleague. "Do not be discontented if you are obliged to devote a large portion of your time to a practical occupation. I also had to do so for a long time and still do. I find that practical tasks protect one from becoming rusty and, contrary to research work, also give one a certain dose of self-esteem, which is so very necessary in life."[15] To the young graduate student Leó Szilárd, Einstein put the question, "Why don't you take a job in the patent office? That would be best for you; it is not a good thing for a scientist to depend on laying golden eggs. When I worked in the patent office, that was my best time of all."[16]

Laying golden eggs was Einstein's frequent metaphor for the often painful duty of a scientist to produce results. He also used eggs, not golden ones, as a metaphor for inventions, sometimes in a confusing manner: for example, in the small poem on a photo (see fig. 4.33) dedicated to one of his inventor partners, Rudolf Goldschmidt. There, he hoped to lay eggs *together* with him.

In his unproductive periods, he re-derived well-known mathematical and physical laws;[17] invented experiments on the character of light, on the mechanism of superconductivity, and on the velocity of gas reactions; and embarked on technological problems both as an expert critic and as a co-inventor. Einstein called his "detours" in technology "escapades"[18] and himself "a frolicsome Sunday rider [the Sunday driver of those times] in the field of technology."[19] These gave him a good feeling of accomplishing something within a reasonable time. Technological problems have well-defined limits. When he heard about a young acquaintance struggling with depression, he recommended a practical occupation, "something like what we had at the patent office," so that he could be occupied with small, well-defined tasks.[20] These "escapades," sometimes substitutes for scientific achievement, gave him satisfaction and, in one case, even a serious temptation to leave his academic career and immerse himself in the everyday problems of an industrial laboratory.

He enjoyed those "escapades," because, as Gerald Holton observed, Einstein had a talent for visualizing both scientific and technological problems.[21] He shared this ability with inventors such as Sperry and Edison.[22] I only mention in passing two of his famous thought experiments: the first, when the teenager Einstein imagined himself riding light waves and looking for what the waves should look like on this peculiar ocean—the germ of special relativity; the other, his clock-in-the-box filled with radiation and with a shutter that emits one photon (i.e., loses an exact amount of energy-mass) at an exact moment—a counterexample to the uncertainty principle. Such thought experiments, however, will not be discussed here.

But we now start the story with Einstein's musings, before moving on to his experiments, his expert opinions, and, finally, his inventions.

CHAPTER ONE

Musings

The Flettner Ship

In February 1925, sensational news swept across the world: a refitted steel vessel was cruising on the Baltic Sea en route from Danzig (the present-day Gdansk, Poland) to Scotland without motors or sails or masts. It was a mystery how the ship, with only two big revolving cylinders resembling smokestacks, designed by German engineer Anton Flettner, cruised twice as fast as when it had 500 square yards of sails (fig. 1.1). In the United States, *Popular Science Monthly* published a report on the vessel, noting that "Prof. Albert Einstein, originator of the theory of relativity, has pronounced the rotor principle of great practical importance."[1]

From March 5 to May 11, 1925, Einstein was on a lecture tour in South America. On March 30, he visited with the editors of the Buenos Aires newspaper *La Prensa* ("an enormous consumption of paper and manpower and technology for ?" he noted in his diary).[2] Apparently on being asked to comment on the new ship propulsion, he prepared an article, which was published in the paper's April 13 issue.[3] In it, Einstein explained that the principle on which rotor propulsion is based (rotors dubbed for the revolving "stacks") had been discovered by Leonhard Euler and Daniel Bernoulli two centuries earlier.

Figure 1.1. Flettner's rotor ship.
Courtesy Library of Congress, George Grantham Bain Collection, ggbain 37764.

Flettner's vessel could be described as a sailing ship in the sense that it is propelled by the wind, but instead of sails, it uses vertical cylinders revolving around their axes. The motion propels the ship by an effect discovered by Heinrich Magnus in 1854.

Figure 1.2 shows a cylinder from above. Suppose that it does not rotate. Then the wind passes both the *A* and *B* sides with the same velocity; no pressure difference appears. As soon as rotation starts, the relative wind speed is increased on the side of the cylinder where their directions are the same (*B*), and reduced on the other side (*A*). According to Bernoulli's law, the higher the speed of a gas, the lower its pressure. So, the pressure along the *A* side is higher than on the *B* side. This excess pressure creates a force that is not *parallel* to the wind (as with traditional sailboats), but *perpendicular* to it, in this case propelling the boat in a direction from *A* to *B*.

The Magnus effect seems obvious, Einstein continued, but chance played a crucial role in its discovery. It was known, but not understood, that cannon

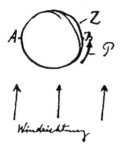

Figure 1.2. The Magnus effect. Windrichtung = wind direction.

Manuscript of Albert Einstein, "El buque de Flettner," *La Prensa* (Buenos Aires), April 13, 1925, p. 10. Courtesy Albert Einstein Archives, The Hebrew University of Jerusalem.

balls change their course even in the absence of wind. The balls being symmetrical, the reason was thought to lie in the ball's rotation around its axis. The same phenomenon—but with tennis balls instead of cannon balls—was observed in England, and the same explanation was given, independent of Magnus, by Lord Rayleigh. Later, Ludwig Prandtl investigated the Magnus effect, both experimentally and theoretically, with a rotating cylinder. Prandtl's work impelled Flettner to make use of the effect in sailing.

Einstein was remarkably well informed about the history of the invention and the roles that Magnus, Rayleigh, and Prandtl played in it. I doubt that he ran to the library of the University of Buenos Aires for a hasty literature search. Rather, he had read the whole story at home, in Berlin. Prandtl published a paper on Flettner's invention in *Die Naturwissenschaften* on February 6, 1925,[4] and the second edition of a booklet on the rotor ship, its physical principles, and the investigations in Prandtl's institute conducted from mid-1923 on the revolving cylinder was published by Prandtl's colleague Jakob Ackeret in the same month.[5] Both Prandtl and Ackeret deliberated on the history of Flettner's invention, mentioning the role that Magnus, Lord Rayleigh, Prandtl, and others played in it.[6] I am sure Einstein read at least Prandtl's paper while perusing this widely read journal before leaving for South America in March, and as an enthusiastic yachtsman, paid particular attention to it.

Flettner claimed that the rotor drive could reduce the crew of ocean liners by two-thirds and fuel consumption by 10 percent. Interest was high both in Germany and in the United States. Already by September 1925, the first American rotor boat was constructed by two MIT students, Joseph Kiernan and W. Hastings,[7] but cheap oil killed its further development. Today, however, in an era of rising oil prices and an increased level of carbon dioxide in the atmosphere, interest in the Flettner rotor has been rekindled. In 1984, C. P. Gilmore informed the readers of *Popular Science* that Lloyd Bergeson built a yacht named *Tracker* with a Flettner rotor to investigate its feasibility.[8] Recent patents on the Magnus rotor, including one granted on December 6, 2007,[9] show that it has not been forgotten.

Why Do Rivers Meander?

"It is common knowledge that streams tend to curve in serpentine shapes," Einstein began his short paper of 1926.[10] "It is also well known to geographers that the rivers of the northern hemisphere tend to erode chiefly on the right side.... Many attempts have been made to explain this phenomenon, and I am not sure whether anything I say in the following pages will be new to the expert.... Nevertheless, having found nobody who was thoroughly familiar with the causal relations involved, I think it is appropriate to give a short qualitative exposition of them."

Stir a cup of tea and watch the leaves swirling at its bottom. Before long, they collect at the center. Why? Because the whirl is hindered by the friction of the bottom and the wall of the cup. The centrifugal force at the bottom will become lower than on the surface, the upper layers of the tea will stream downward along the wall, and the leaves at the bottom will be swept into the center. Then the stream rises from the center of the bottom as a fountain. Nowadays this flow is called secondary or helical flow (fig. 1.3).

The same phenomenon happens with a stream at its bends. A centrifugal force acts on the outer side of the bend, but this force is lower near the bottom because the flow is hindered by friction with the bottom. As a consequence, a helical flow will develop (fig. 1.4). This flow develops even where there is no bend, because the Coriolis force, acting transversely to the stream, diminishes

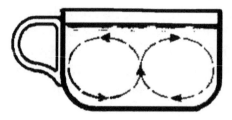

Figure 1.3. Flow in a teapot.
Albert Einstein, "Die Ursache der Mäander-bildung der Flußläufe und des sogenannten Baerschen Gesetzes," *Die Naturwissen-schaften* 14 (1926): 223–24. Courtesy Albert Einstein Archives, The Hebrew University of Jerusalem.

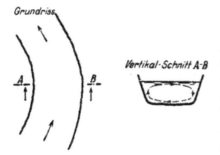

Figure 1.4. Helical flow in river bed. Grundriss = ground plan; Vertikal-Schnitt A–B = vertical section A–B. Albert Einstein, "Die Ursache der Mäanderbildung der Flußläufe und des sogenannten Baerschen Gesetzes," *Die Naturwissenschaften* 14 (1926): 223–24. Courtesy Albert Einstein Archives, The Hebrew University of Jerusalem.

with depth. These are the preliminary considerations with which Einstein introduced his paper.

The fact that right banks of northward flowing rivers in the northern hemisphere have the tendency to erode more strongly than the left ones was observed by several scientists in the nineteenth century, among them Karl von Baer. It is called Baer's law after him. Einstein did not go into detail, but he saw a problem: the observed right-side erosion must be reconciled with the fact that when a straight stream starts to bend, the water "rushes out" to the outer bank under the influence of the centrifugal force and erodes it faster than the inner side, regardless of whether it is a left or a right bank. What is more, the hydraulic principle suggests that when a straight stream flows into a curve, the water must slow down along the outer side and flow faster on the inner side, so that erosion must be stronger along the *inner* side.[11]

Einstein then considered the velocity distribution of the stream over its cross section. Suppose that still water is uniformly accelerated. At first, if the velocity distribution is uniform, the banks and the bottom will slow the neighboring layers, and by internal friction a steady state is reached. Whirls developing along the banks are swept toward the center, where they dissolve. Equilibrium is reached slowly, so that even small disturbances can influence the eventual stream pattern. The helical flow, though a tiny disturbance, makes itself felt. It carries the faster streamlines on the surface from the left bank to the right and the slower ones close to the bottom from the right to the left bank. The helical flow carries the material of the right bank toward the left one.

Another consequence is that the helical flow, due to its inertia, reaches its maximum somewhat after the bend, so meander formation continues. A third consequence is that if the river's cross section is large, it takes more time for the friction at the banks and bottom and the viscosity of the water to slow down the stream, which is why larger rivers make meanders of longer wavelength.

The paper was read not only in Germany but also in the United States. John R. Freeman, of Providence, Rhode Island (probably John Ripley Freeman, a metallurgist), was eager to check whether Einstein's considerations were known to specialists in the field. He found an article in the *Proceedings of the Royal Society of London* from 1876 on the causes of meandering of rivers.[12] He forwarded this information to a German pen pal, who was also a pen pal to Ludwig Prandtl. Prandtl communicated it to the journal *Die Naturwissenschaften,* in which Einstein's paper had been published a few months earlier.[13] To show that a meticulous search of the German literature would have revealed what was original in Einstein's paper and what was already well known before 1926, Prandtl unearthed two papers from 1896 and 1911, admitting that the first was not easily accessible.[14] He also listed a few publications on experimental findings and added a detail of meander formation not mentioned by Einstein.

Prandtl was right. In 1896 James Isaachsen discussed the "secondary motion" (the helical flow) in a vessel and in water streams.

In 1943 Max Born asked Einstein whether he had expanded upon his ideas on meandering in more detail somewhere else.[15] "My remark about the influence of the curvature of rivers and of the Coriolis force on the erosion of waterways was only a casual one," Einstein answered. He was convinced, he added, that the idea must have been known for a long time. "I have, however, never searched the literature."[16]

Einstein's contribution was forgotten. The curious behavior of tea leaves caused a whirl in the *American Journal of Physics* in 1951 and 1959,[17] where the phenomenon was discussed with no mention of Einstein's contribution. R. A. Alpher and R. Herman did justice to him in 1960.[18] They also found that, among the precursors, Harold Mottsmith and Irving Langmuir had already discussed the helical flow in colloidal solutions[19] without being aware of Thomson's results. "Even a generation ago it was difficult to ensure that one had seen all the literature on a given problem," Alpher and Herman concluded.

Not that our contemporaries were better in seeing "all the literature on a given problem." After Alphen and Herman had discovered Einstein the geologist in 1960, in 1988 Kent A. Bowker made a search of the literature,[20] and he found mention of Einstein's name only in S. A. Schumm's short comment;[21]

Andrew S. Goudie, in his paper on the one-and-a-half century long dispute over Baer's law,[22] found a reference to Einstein in A. E. Scheidegger's book on geomorphology.[23] Jesus Martinez-Frias et al. in 2005 also complained of the lack of references to Einstein.[24] They found the earliest satisfactory tribute to Einstein in two papers from 1995 and 2001.[25] None of them mentioned Alpher and Herman's earlier discovery. True, the papers discussing the swirling of tea leaves in the *American Journal of Physics* are short notes, but they merited a meticulous search.

Since then, Einstein's name has been tied to helical flow, at least in the American literature on the history of science. In *ABC Science Online* news from January 17, 2007, Stephen Pincock announced that "Einstein's tea-leaves inspire new gadget."[26] The inventor, Leslie Yeo, is now working on a credit-card-size diagnostic kit.[27] A blood sample is placed in a tiny container. By pointing a needle at an angle across the surface of the liquid and by applying a voltage to the needle, a wind of ions is produced that sets the blood sample in circulation. The blood cells gather in the center at the bottom, just like the tea leaves, where they can be analyzed.

Einstein's explanation of the cup-of-tea experiment has passed the test of time. Meandering, in contrast, has proved to be far more complex than Baer's law and helical flow suggest. Fluvial geomorphology has since established that the Coriolis force is negligible in rivers that are smaller than ocean currents or the atmosphere. The debate on the importance of the Earth's rotation in geology continues. A. S. Goudie drew the conclusion that "whether what is a modest force, operating continually and over an extended period of time, can achieve major landscape change, is central to the debate. That said, no fully convincing renunciation of Baer's law has yet been produced."[28]

Einstein presented his considerations to the Prussian Academy of Sciences on January 7, 1926, with the title (which also served for an abstract) "On the Cause of Meander Phenomenon of Courses of Rivers (Influence on the Mean Velocity at the Bank by a Circulation Caused by a Locally Varying Centrifugal Force)." He immediately continued with another lecture "On the Application of a Split of the Riemannian Tensor of Curvature, Found by Rainich, in the Theory of Gravitational Field." Against the background of this second lecture, the first looked like, or was intended to be, a conjurer's trick.

The meander lecture was published in the March 12 issue of *Die Naturwissenschaften*, which was dedicated to the eightieth birthday of Emil Warburg,

Einstein's colleague. Warburg was a member of the physical-mathematical class of the Prussian Academy and may have been present when Einstein delivered his lectures. Perhaps Warburg was impressed by the first so much that Einstein offered its published form to him as a birthday present.

CHAPTER TWO

Experiments

Michelson, Morley, and Eötvös Reinvented

According to Einstein, "The theoretically oriented scientist cannot be envied because nature, i.e. the experiment, is a relentless and not very friendly judge of his work. In the best case scenario it says only 'maybe' to a theory, but never 'yes' and in most cases 'no.' If an experiment agrees with theory it means 'perhaps' for the latter. If it does not agree it means 'no.' Almost any theory will experience a 'no' at one point in time—most theories very soon after they have been developed."[1] Following his own maxim, Einstein often turned to nature and requested its judgment by proposing, and performing, experiments.

The Michelson-Morley and the Eötvös experiments were critical to Einstein's work and form the pillars of his most famous theories, the special and general theory of relativity. One might think Einstein constructed his theories by first having learned of the experiments and then erecting his theories upon these foundations, but it did not happen that way. He did not bother much about digging up literature (even if he maintained the opposite) but blazed his own trail. As he reported in 1922,

the first time I entertained the idea of the principle of relativity was some seventeen years ago. . . . Light travels through the ocean of the ether, and so does the earth. From the earth's perspective, the ether is flowing against the earth. And yet I could never find proof of the ether's flow in any of the physics publications. This made me want to find any way possible to prove the ether's flow against the earth due to the earth's motion. . . . Thus I predicted that if light from some source were appropriately reflected off a mirror, it should have a different energy depending on whether it moves in the direction of the earth's movement, or in the opposite direction. Using two thermoelectric piles, I tried to verify this by measuring the difference in the amount of heat generated in each.[2]

Einstein says he did not find decisive information that proved the ether flow, which means he did not find the Michelson-Morley experiment either, which was performed in 1886–87. Or so he remembered in 1922. For us, it is important that he set up an experiment to detect the Earth's velocity with respect to the ether. Alas, no further details are known because he did not publish anything about this experiment.

But he had thought about the ether earlier. He wrote to his fiancé, Mileva Marić, in 1899 about "a way of investigating how the relative motion of bodies with respect to the luminiferous ether affects the velocity of propagation of light in transparent bodies" and that "I . . . wrote to Professor Wien . . . about the paper on the relative motion of the luminiferous ether against ponderable matter."[3]

The case with general relativity is no less surprising.

Einstein wrote to Wilhelm Wien (again) in 1912 regarding inertial mass. Suppose we have a mass of uranium that decays radioactively to form lead. The *inertial* mass of the lead is less than that of uranium, because of the α-radiation the uranium has emitted. But an exact balance exists for the *gravitational* mass, so that the relative difference between the oscillation periods of a uranium pendulum and a lead pendulum in the same gravitational field "should be about 2×10^{-4}, which could be easily proved."

Einstein added a postscript describing a more sensitive method than a pendulum: "A torsion balance with a piece of uranium and a piece of lead placed on its beam must experience a torque when the beam is brought into the west-east direction, with the torque changing its sign when the balance is rotated by 180°. As I established through calculation, this effect should be quite easy to measure. Would you, perhaps, be so kind as to have this simple experiment—which would have the significance of an *experimentum crucis*—carried out?"[4]

No reply from Wien is extant, so we may only guess that Wien informed Einstein that the proposed setup was exactly Roland Eötvös's torsion balance with which the Hungarian geophysicist established the proportionality of inertial and gravitational mass with a precision of 5×10^{-6} —in 1889.[5]

The Mass of the Electron

Einstein's next proposal for an experiment fit in the mainstream of the theory of electrons. Its father, the great Dutch physicist Hendrik A. Lorentz, established that an electron, moving in an electromagnetic field, has two kinds of mass: a "longitudinal" and a "transversal." In other words, the mass (namely the ratio of the force to the acceleration it gives to a body) is different in the direction of the electron's motion and perpendicular to it. In Einstein's formulation,

$$\frac{\mu}{\left(\sqrt{1-\left(\frac{v}{V}\right)^2}\right)^3}$$

is the longitudinal, and

$$\frac{\mu}{1-\left(\frac{v}{V}\right)^2}$$

is the transversal mass. μ is the mass of the electron at rest, v is its velocity, and V is the velocity of light.[6]

This peculiar dependence of mass on velocity was a novelty that had to be confronted by experiment, especially because Lorentz's model of the electron was not the only one at the time. Other, competing theories arrived at different expressions.

The first attempt at an experimental test was made by Walter Kaufmann,[7] whose data contradicted Lorentz's (and Einstein's) theories. Einstein set forth to find the dependence of the ratio of the transversal mass to the longitudinal mass on the potential that accelerates the electrons in a cathode ray:[8] namely, the higher the potential (i.e., the higher the velocity of the electron), then the smaller this ratio than 1. Figure 2.1 is a sketch of the experimental arrangement he proposed in 1906.

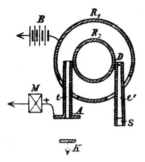

Figure 2.1. Measurement of the electron's charge.
Albert Einstein, "Über eine Methode zur Bestimmung des Verhältnisses der transversalen und longitudinalen Masses des Elektrons," *Annalen der Physik* 21 (1906): 3–56. Courtesy Albert Einstein Archives, The Hebrew University of Jerusalem.

The electrons are emitted by the grounded K cathode and accelerated by the potential between K and A. M is a power supply. The electrons enter the space between metal cylinders R_1 and R_2 with a definite v velocity and are collimated by the tube t. R_1 is grounded, and R_2 and t are connected to the positive pole of the supply. The voltage difference between the two cylinders creates a radial force on the electrons. If correctly set, the electrons will follow a semicircular orbit until they reach tube t', which has fluorescent screen S at its end. The shadow of the vertical wire D is projected on the screen.

As we increase the potential, the electrons' masses will change as their velocity increases; they will follow a different orbit, and the shadow of D on the screen will shift from its original position. To restore the shadow to its original place, an additional potential is used from an auxiliary battery B. This way the dependence of the ratio of the masses on the potential can be established and compared with the predictions of the competing three theories, namely:

$$\frac{\mu_t}{\mu_l} = 1 - 0.0070 \frac{\Pi}{10000}$$ according to Alfred Bucherer,

$$\frac{\mu_t}{\mu_l} = 1 - 0.0084 \frac{\Pi}{10000}$$ according to Max Abraham, and

$$\frac{\mu_t}{\mu_l} = 1 - 0.0104 \frac{\Pi}{10000}$$ according to Lorentz and Einstein,

where Π is the potential in volts.

Einstein confessed that he himself was not in a position to carry out the experiment, but he offered the idea to interested physicists. An experiment following Einstein's idea was never performed. The competition between electron theories ended with the triumph of the Lorentz-Einstein theory in 1916–17, via spectroscopy.

Ampère's Molecular Currents

In 1915 Einstein himself performed an experiment, in collaboration with Wander de Haas, a Dutch colleague who was also Hendrik A. Lorentz's son-in-law.[9]

Einstein's goal was unusual, for it dealt with neither relativity nor quantum theory. He wanted to check an almost century-old hypothesis.

In 1820 André-Marie Ampère proposed that the magnetic properties of substances are due to tiny electric circuits in their molecules. If, according to Lorentz's electron theory, we add that these molecular currents, as every electric current, consist of moving elementary charges, electrons with inertial masses, then Ampère's hypothesis is equivalent to saying that there are microscopic gyroscopes in ferro- and paramagnetic substances. If their axes are set parallel and they all rotate in the same direction, their magnetic fields add up to a macroscopic magnetic field: the substance gets magnetized.

But there were arguments against this hypothesis, Einstein and De Haas continued. According to classical electrodynamics, revolving electric charges must radiate, so these molecular electric circuits must lose energy and finally collapse. Furthermore, the Curie-Langevin law stipulates that the magnetic moment of molecules is independent of temperature, so that the electrons must revolve even at absolute zero degree. If Ampère was correct, there must be a zero-point energy, an idea that many physicists were reluctant to accept.

After setting the scene, the authors formulated the concept of the experiment. The revolving electron ε as a gyroscope has an angular momentum and a magnetic moment parallel to it (fig. 2.2). If we suspend a rod and change its magnetization, the angular momentums of its elementary circuits must also change. As momentum is conserved, the macroscopic angular momentum of the rod must counterbalance this change. A torque would prove the existence of molecular currents.

Einstein and De Haas suspended an iron rod S from a glass fiber G at the center of two electric coils. The fiber is attached to a brass tube E. The effective length of G (with it the proper torsional frequency of S) is controlled by screw P, which can change the vertical position of clamp B that holds G. Two mirrors are fastened back to back to S at the gap between the coils. The mirrors reflect a beam of light that serves as an indicator of the twisting of the fiber (fig. 2.3). The authors claimed to have eliminated the disturbances due to the magnetic field of the Earth, to a permanent magnetization of the rod, to eddy currents in it, and so on.

To begin, the rod was magnetized axially. The current in the coils was reversed, and this change in the magnetic field was expected to generate a

Fig. 1.

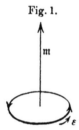

Figure 2.2. The magnetic field of a rotating electron.
Albert Einstein and Wander J. de Haas, "Experimenteller Nachweis der Ampèreschen Molekularströme," *Deutsche Physikalische Gesellschaft, Verhandlungen* 17 (1915): 152–70. Courtesy Albert Einstein Archives, The Hebrew University of Jerusalem.

Fig. 4.

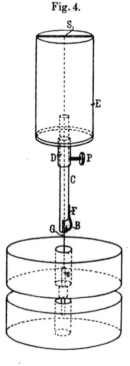

Figure 2.3. Experimental setup for Ampère's molecular currents.
Albert Einstein, and Wander J. de Haas, "Experimenteller Nachweis der Ampèreschen Molekularströme," *Deutsche Physikalische Gesellschaft, Verhandlungen* 17 (1915): 152–70. Courtesy Albert Einstein Archives, The Hebrew University of Jerusalem.

torque. To amplify an effect, they switched the direction of the current at the same frequency as the natural torsional frequency of the rod, which sets up a resonance effect.

The experiment gave a positive result. The authors were even convinced that they got close to the theoretical result calculated with the mass and charge of the electron, thus proving not only that Ampère's molecular currents exist but that they consist of circulating electrons.

A week after the publication of their experiment, Einstein described a simplified version to use in lectures. He performed it together with two assistants from the Physical-Technological Establishment (Physikalisch-Technische Reichsanstalt),[10] but this new arrangement disclosed a serious problem: the magnetic forces in play were far greater than the effect to be measured. To keep them at bay, Einstein dropped the resonance method and turned to a sudden change in the electric current, a discharge of around 0.001 second, just enough to reverse the remanent magnetism of the rod. To amplify the twist, he set the rod swinging by means of a small permanent magnet and discharged a condenser whenever the rod reached the peak of its swing.

Similar experiments carried out before, simultaneously, and after Einstein and De Haas's experiment cast doubt on their results. De Haas waged a rearguard fight up to 1921 and attempted to eliminate possible sources of errors. In the end, it turned out that their result,

$$M = \frac{2\mu}{\varepsilon} J$$

did not stand the test of time. Here M is the magnetic moment of the rod, J its angular momentum, and μ and ε are the mass and the charge of the electron, respectively.

Why did Einstein spend so much time in a field far from his main interest? Peter Galison argues that the (at least qualitatively) positive result was a lifeline for Einstein's sinking faith in the existence of zero-point energy, a problem he had failed to solve theoretically since 1912. As such, it was not really that distant from his interest in quantum theory.[11] Galison places the Ampère venture in the broader context of Einstein's methodology: it was stimulated by his striving for unification. Ampère had proposed molecular currents to find a common cause for permanent and electrically induced magnetism. Einstein's experiment demonstrated it qualitatively. Adding Lorentz's hypothesis, that electric currents consist of electrons, coupled the electron theory to Ampère's "unification." Finally, it turned out that Bohr's already well-known hypothesis—that electrons circulate around the atomic nucleus without radiation and do it even at absolute zero (i.e., they must have a zero-point energy)—comprised statements that are no longer independent but are somehow related to each other, however obscure and weird this relationship may be.

There is a remarkable confession in an Einstein letter: "It was an expert opinion which I prepared [for a patent process] on a gyrocompass . . . that inspired me to prove the gyroscopic nature of the paramagnetic atom."[12] With eyes sharpened by this statement, we can discover a trace of it in the first paper: "The magnetic molecule behaves like a gyroscope," and "this is an angular momentum of the same kind as the angular momentum in the theory of gyroscopes."[13]

This illustrious expert opinion is discussed in chapter 3.

Later, in an undated letter, Einstein proposed to De Haas an interesting topic to investigate, "which is even connected to our previous work (the gyroscopic effect of elementary magnets)," namely to check whether a magnetomotive force appears when a ferromagnetic rod is mechanically rotated.[14] The same effect should be found when only the elementary magnets in the rod are rotated by a magnetic field.

Two Americans hunted for the effect, he continued, but they did not succeed in finding it. Einstein was of the opinion that the strength of the rotating magnetic field they used was not strong enough to make the rod's magnetization saturated (i.e., to align the elementary magnets close to parallel); consequently the magnets did not follow the rotating field but swung to and fro.

As a remedy, Einstein proposed to magnetize a piece of iron in the X direction close to its saturation and then to apply a rotating magnetic field in the Y direction. As a consequence, a magnetomotive force should appear in the Z direction, which can be detected by the electric current it induces (fig. 2.4).

He proposed the experimental arrangement shown in figure 2.5. The iron ring consists of lamellas. The induced magnetomotive force was expected to be observed in the direction of the ring's radius.

There are no further data on the fate of this idea. I must add, however, something about the time this letter may have been written. A hint at the date would come from finding the two Americans he mentioned. I succeeded only in finding a British physicist, J. W. Fisher, who published on exactly this experiment, first in 1922, and again in 1925, and announced negative results.[15] A German-language referate of Fisher's 1925 paper was published in the *Physi-*

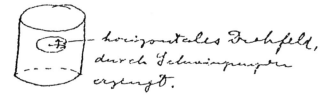

Figure 2.4. Principle of the experiment. horizontales Drehfeld durch Schwingungen erzeugt = horizontal rotating [magnetic] field produced by swingings.

Einstein to Wander de Haas, [1922–28]. Courtesy Albert Einstein Archives, The Hebrew University of Jerusalem.

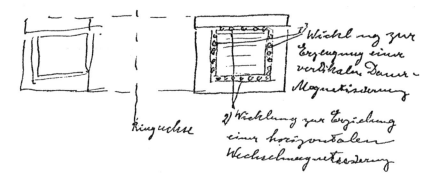

Figure 2.5. Experimental setup. 1) Wicklung zur Erzeugung einer vertikalen Dauer-Magnetisierung = winding for generation of a vertical permanent magnetization; 2) Wicklung zur Erzielung einer horizontalen Wechselmagnetisierung = winding to produce a horizontal alternating magnetization; Ringachse = ring axis.

Einstein to Wander de Haas, [1922–28]. Courtesy Albert Einstein Archives, The Hebrew University of Jerusalem.

kalische Berichte in mid-1927.[16] By supposing that Einstein might have learned about Fisher's experiments in it, all these add up to a time span of 1922–28 for Einstein's letter.

In a letter of August 1924, Peter Pringsheim reported on a similar experiment proposed by Einstein. He suspended a steel sphere on a thin iron wire and observed that, when magnetized and swung in the north-south direction, the sphere began to rotate. This he called "the supposed effect." When the sphere swung in an east-west direction, the rotation turned stronger, which Pringsheim considered to contradict the first effect. When he put a small magnetometer near the sphere, the rotation stopped. What he so achieved he did not call a proof of the expected effect but "a probably very sensitive method to detect weak inhomogeneities in the terrestrial magnetic field."[17] I cannot make a complete picture out of these fragments of what Einstein's "supposed effect" was.

The Velocity of Gas Reactions

Einstein began a short paper in April 1920, "Although our knowledge of the chemical equilibrium of gases is well advanced,we have only inadequate knowledge about the reaction velocity of gas reactions."[18] And with good reason, because the catalytic activity of the rigid wall surrounding the gas (in other words, how much the wall aids or hinders the reaction), the high temperature, and the high values of reaction velocities all make the measurement difficult.

To overcome these difficulties, Einstein proposed to determine this velocity by measuring sound propagation in a mixture of dissociated and non-dissociated gas molecules.

First, send a low-frequency sound through such a mixture. The sound wave alternately compresses and rarefies the gas, but the local volume changes will be slow enough for the chemical reactions to take place through states of chemical equilibrium. Now raise the frequency. At a certain value, the local volume changes become adiabatic, no chemical changes can take place, and so the gas will behave as a normal mixture does, and its compressibility decreases. Consequently, when we raise the frequency of the sound, there is no change in its velocity at first; then we pass a point after which the velocity will increase, until we reach a frequency where the velocity again becomes constant. Between these two values, the duration of the reaction is shorter than, or equal to, the period of the sound.

As Einstein remarked, Friedrich Keutel had established in 1910 that the sound velocity depends on the reaction velocity of the gas through which it propagates and that from this dependence one can determine when the dissociation reaches the state of equilibrium.[19] Einstein went down Keutel's road further by calculating the constant of velocity of dissociation from the change of sound velocity with its frequency.

At Einstein's suggestion, Eduard Grüneisen and Erich Goens performed experiments with N_2O_4 gas to check Einstein's prediction.[20] They could not go above 15,600 cycles and did not yet reach the lower limit of the range at which the sound velocity was expected to change. This indicated that this gas dissociates in less than 1/15,600 seconds.

As a result of his paper, Einstein has been ranked among the pioneers of molecular acoustics.[21]

A Geodynamo Model?

In 1921, on the basis of a proposal by Einstein, Karl Glitscher and Max Schuler, both engineers in Hermann Anschütz-Kaempfe's gyrocompass factory in Kiel, tried to find out whether a rotating heated body gives rise to a magnetic field.

They heated a copper cylinder first by friction then by circulating hot oil through it. An exchange of letters between Einstein and Schuler indicates that they did not agree on how to measure the expected magnetic field. Einstein stuck to a simple magnetic needle to get a qualitative result, whereas Schuler and Anschütz preferred to use a coil around the cylinder to amplify the induced electric current and thereby increase the sensitivity.[22]

Because of the lack of details, the purpose of this experiment can only be guessed.

Bernhardt Schell suggests it was part of the development of the gyrocompass. It probably "served to check the influence of eddy current fields which might occur on the inside of a sphere. It was necessary to determine what forces were generated by the rotating gyros that could influence the north stability."[23]

Einstein's remarks, scattered throughout his correspondence, can be interpreted in another way. "Even though I cannot yet form a clear picture of whether a positive effect is to be expected," he wrote to Schuler, "it is still, for me, the only plausible possibility to combine the *heat current* and the earth currents, since the latter can only be caused by an *irreversible* process"—a rather vague formulation.[24] What exactly do "heat current" and "earth currents" mean? Did he have in mind an analogy between the experiment and a process taking place within the Earth? At another place he wrote to Anschütz that a positive result "would be of enormous interest."[25] Enormous interest for whom? Would he have used the word "enormous" if he had had in mind only the small group working on the gyrocompass? In addition, in another letter he called the experiment "geomagnetic,"[26] and in his reply to the news that Schuler found a negative result, he admitted that "in thinking about the nature of the Earth's magnetic field I have got bogged down in improbabilities."[27]

Was Einstein curious about the origin of terrestrial magnetism, and did he guess that its cause might be the rotation of the high-temperature inner mass of the Earth? I am inclined to say yes. The question of why the axis of rotation and the magnetic axis of the Earth almost coincide had occupied him already in 1915 when he drew the conclusion that, with the positive result of the experiment on Ampère's molecular currents, "the reason has also been found why the Earth's magnetic axis and axis of rotation almost coincide."[28]

Did he lose his confidence in this conclusion? Or did his search for a connection between acceleration and magnetism have its roots in his search for a unified field theory? "I'd be happy to learn in which direction you chivvied the problem of the unification of physical fields that you often talked with me about," Hans Mühsam wrote to him in 1942. "Do you remember," he continued, "that in connection with this question you traveled on the commuter train to Lichterfelde with a small pocket compass to check whether the mechanical acceleration of the railroad gives rise to a field corresponding to a magnetic field? The expected effect could not be found because of the insufficiency of the device."[29]

He returned to the problem in 1923, together with Hermann Mark, a physical chemist. They even started a paper with the surprising title, "On an Obvious Hypothesis on the Cause of the Geomagnetic Field and Its Experimental Refutation." Unfortunately, we have only the first page of the manuscript in Einstein's and Mark's hands;[30] according to Mark's later memories, it consisted of three or four pages.[31] The available page sets forth the very first step in a theoretical consideration that has its roots in Einstein's new theory, an early attempt at a unified theory of gravitation and electromagnetism.[32] Einstein and Mark consider an equation for the components of the electromagnetic four-potential and maintain that this equation "leads to a simple explanation of the part of the electromagnetic field, the axis of which coincides with the axis of rotation of the Earth." Obviously, the hypothesis mentioned in the title of the manuscript must have been connected with this theory. The experiment proposed to refute it was not indicated.

The experiments led to no conclusive results, and the manuscript was never published.[33]

Five years later, W. F. G. Swann and A. Longacre intended to explain the Earth's magnetism by its rotation.[34] They rotated a copper sphere of 10 centimeters in radius with a speed of 200 rotations per seconds (without heating it). Then they compared the Earth's magnetic field and that of the sun, taking into consideration their rotation; applied the results to the sphere and the Earth; and calculated the sphere's expected magnetic field. The strength of the magnetic field they observed was less than 7×10^{-4} gauss, which they did not accept as an evidence. I mention this attempt only to add weight to the guess that with the rotating hot cylinder Einstein tried to find the source of the terrestrial magnetic field.

Light: Waves or Particles?

The first half of the 1920s was one of the most dramatic periods in the history of physics: that of the parturition of quantum mechanics. Einstein's proposal of the energy quantum in 1905 was being taken more and more seriously, but far from being accepted in general. The photoelectric effect was the only phenomenon that can be explained with it; the vast majority of optics, eminently the interference of light, contradicted it. Niels Bohr was a staunch supporter of the wave theory of light; Arnold Sommerfeld seemed to wait for a decisive answer. Einstein looked as though he might favor corpuscles, even though he formulated his proposals impartially—as attempts at a decision between the two competing viewpoints.

He also approached the problem by sketching a field theory that promised to derive quanta as singularities of a field, a unified one, comprising both general relativity and electromagnetism.

In addition to his theoretical efforts, perhaps because of his impatience or impotence, Einstein tried to untie this Gordian knot by cutting it with the sword of experiments.

"Can we decide by experiment whether the electric field in a radiation really follows the [energy] distribution demanded by Maxwell's theory?" he asked Lorentz on the very first day of 1921. "Answer: according to the theory, in strong radiation fields the electric fields are of the magnitude of 150 volts per centimeter. These should lead to an observable [spectral] line broadening, due to the Stark effect. We intend here [in Berlin] to carry out experiments on it, and also to browse through astronomical data. . . . The line broadening must grow with the number of order of the line."[35] He used the plural "we" because he had cooperated with Peter Pringsheim, an experimental physicist at the University of Berlin.[36]

He mentioned astronomical data because stars are very hot gas spheres—that is, a very strong radiation field prevails in them, exceeding any field that can be produced in a laboratory—so a broadening must be observable in their spectral lines. Today the Stark broadening of plasma is a special topic of research in astronomy. Was Einstein (with Pringsheim) the first to turn attention to it?

In addition to Lorentz, he mentioned his plan to Max Born, and now he went into detail.[37] Consider a strong thermal radiation with a mean field intensity of 100 volts per centimeter. If energy follows Maxwellian distribution, a Stark effect must be observed in both the emission and the absorption spectra. If, however, the energy distribution follows the statistical law (which considers radiation as a photon gas), the Stark effect will be concentrated on a few molecules and will be very strong, so that near to sharp spectral lines, very weak diffuse lines will appear. "I want to test out this matter with Pringsheim; it isn't easy." (I apologize for using the term "photon" for the energy quantum when talking about events before 1926 when this term was invented.)

For Born, the idea was "bold" and "very nice";[38] for Paul Ehrenfest, with whom he also shared his plans, it was a "great thing," but sounded "unbelievable."[39]

Apparently unsatisfied with the course of the experiment, in August 1921 he launched a second venture. "I thought up a very interesting and quite simple experiment on the nature of light emission. I hope I can conduct it soon," he wrote to Max Born.[40] He went into detail in his letter to Arnold Sommerfeld: "I am doing an interesting experiment with Geiger on the light emitted by a canal ray particle. Problem: Is the interference field generated by a canal-ray particle moving in the direction of the arrow in an elementary process really bluer at *A* than at *B*? If yes, the light ray has to be bent by a dispersive medium."[41] (For details, see below.) He had great expectations and recounted the experiment to both Heinrich Zangger and Michele Besso.[42]

He submitted a paper on his plan in December.[43] In it, he considered gas molecules in a discharge tube filled with a rarefied gas. Cathode rays (electrons), flying from the negative electrode toward the positive one, ionize the gas molecules. These positive ions (canal rays) move toward the cathode and are absorbed there. They emit light when hit by surrounding gas molecules. The proposed experiment is sketched in figure 2.6.

If light consists of waves, the emitted light will propagate spherically. Because of the Doppler effect, their frequencies will be different along different directions. When the illuminating particles of the canal rays move upward

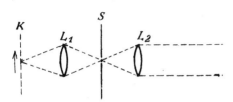

Figure 2.6. Checking the nature of light. Albert Einstein, "Über ein den Elementarprozeß der Lichtemission betreffendes Experiment," *Preußische Akademie der Wissenschaften* (Berlin), *Sitzungsberichte* (1921): 882–83. Courtesy Albert Einstein Archives, The Hebrew University of Jerusalem.

along K, the frequencies of the light waves reaching lens L_1 at its lower rim will be lower, and they pass through the slit S to the upper rim of lens L_2. In contrast, the frequency of waves that finally reach the lower rim of L_2 will be higher. Because of the difference in their wavelengths, the surfaces of equal phases of elementary waves will not be parallel to each other but will form fans; the average direction of propagation, however, will not change.

Put a layer of dispersive medium in the way of the light. Now the elementary waves with higher frequencies will propagate more slowly than those with lower frequencies, so that the light beam will be deflected. According to a simple calculation, this deflection can easily be observed.

If, however, light is not a wave but consists of particles, then the elementary emission should have only one frequency, defined by the quantum of action $h\nu$, and there will be no deflection, even when the source is moving. This had to be reconciled with Johannes Stark's observation of the Doppler shift of canal rays, but Einstein did not consider this problem insurmountable[44] and announced that he had already started setting up the experiment with Hans Geiger.

We have two interim reports on the experiment. On November 7, Geiger informed Einstein that he did not find deflection so far. In his arrangement, canal rays run along a metal tube with a diameter of 8 millimeters and filled with carbon disulphide. The expected deflection after a path of 50 centimeters would amount to 50 graduation in the ocular.[45]

A month later, Walther Bothe thought he had found a mistake in Einstein's preliminary calculations, and discussed it with him by the phone.[46]

Finally, the experiment gave a negative result, that is, it seemed to confirm the particle nature of light.[47] As Einstein announced to Max and Hedwig Born with exultation, "It is surely proven that the undulatory field has no real existence and that Bohr's emission is an instantaneous process in the real sense. It is my most powerful scientific experience in years."[48] When Arnold Sommerfeld learned about the news, he remarked: "So you made another great discovery, buried the wave theory. . . . I am glad if you can get a spy hole into some tip of it. The way things have been with the dualism in viewpoints cannot go on."[49]

Neither Ehrenfest[50] nor Laue[51] accepted this conclusion, and both argued that even classical wave theory would have given a negative result. Einstein admitted that he was wrong,[52] corrected his derivation,[53] and, indeed, got a zero deflection even with classical wave theory. "The experiment on which I had placed so much importance proves nothing for and nothing against the undulatory theory, so all the labors of love were actually in vain," he wrote to his sons.[54]

He did, however, not give up. The same day he announced the failure to his sons, he sent a sketch of another experiment to Paul Ehrenfest (fig. 2.7). "Cathode rays K that fall on one of two parallel planes of the same thickness very obliquely produce planar light at A. The plates generate interference without phase difference—provided emission proceeds in a spherical wave. Has a similar experiment been performed somewhere? A positive result is very probable, however."[55] No answer by Ehrenfest is known.

He had another idea in mind already in January, namely to determine, with Geiger and Bothe, the quadratic Doppler effect "by means of a few little tricks; but it is difficult."[56] Again, he announced no further news about the tricks. In 1923, however, the idea emerges again in a letter by W. Orthmann to Peter Pringsheim,[57] in which he proved by calculation that neither the simple nor a higher-order Doppler effect can be detected. Pringsheim forwarded the letter to Einstein with the remark that the calculation underlines what he himself had expressed to Einstein in words.

The cooperation with Pringsheim did not stop either in these years. Around mid-July of 1922, the problem to be solved was the same as that of Geiger and Bothe: to determine whether emission takes place instantaneously or in a longer time. As Pringsheim reported to Einstein, the experiment failed again. He offered two explanations why spectral lines broaden even in the case of momentary emission.[58] Again, no further details are available.

Einstein was adamant. In 1923 he sketched a "fundamental experiment," as Pringsheim called it,[59] but no details are known.

Another glim of hope to get rid of the dual nature of light arose in August 1923. Arnold Sommerfeld, who had just returned from the United States,[60] informed Einstein about Arthur Compton's brand-new theory on the scattering of X-rays on electrons.[61]

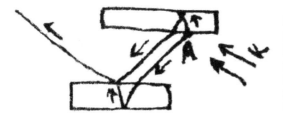

Figure 2.7. Another proposal for checking the nature of light.

Einstein to Paul Ehrenfest, February 12, 1922. Courtesy Albert Einstein Archives, The Hebrew University of Jerusalem.

When light elements were irradiated by X-rays, the wavelength of scattered beams was longer than that of the primary beam, indicating that energy was transmitted to the scattering electrons. This was hard to explain with electron theory. According to that theory, the incident radiation sets the electrons in the matter in vibration at the same frequency. The vibrating electrons also send out electromagnetic waves, also at the same frequency as their vibration. The hypothesis that the cause of scattering and dispersion of light by matter is due to the interference of these waves accounted for these phenomena rather well, but it was unable to explain why scattered radiation had a longer wavelength. Compton supposed that the energy of the primary radiation was transmitted to the electrons by collision of individual X-ray photons with individual electrons— that is, he made use of Einstein's photon hypothesis of 1905. Einstein proposed the concept of quantum of radiation for explaining a particular phenomenon: the photoelectric effect, he remarked. "In order to carry any great weight, the [photon] hypothesis should also be found applicable to phenomena of widely different character. . . . The change in wavelength of the scattered rays, and the recoil electrons associated with them, . . . are just such phenomena."[62]

In formulating the theory, Compton relied not only on earlier experimental findings but also on his own.

Einstein was electrified by Compton's result because what Compton had found was no less than evidence for the corpuscular nature of X-rays and a strong argument for the existence of photons. He even wrote a long article on the importance of Compton's experiments to the daily newspaper *Berliner Tageblatt.*[63]

Einstein wanted to repeat Compton's experiments. He turned again to Hermann Mark, whom we met in the previous chapter. Mark set about securing accessories for the experiment, including a Coolidge X-ray tube with rhodium anticathode, rhodium, and molybdenum foils, and he made a search in the literature.[64] He discovered that in March 1923 Peter Debye had explained the same experimental findings with quantum theory and published his account in the April 15 issue of the *Physikalische Zeitschrift*.[65] As to the strength of experimental evidence for Compton's theory, Mark also found a short paper in *Nature* in which the authors reviewed experiments on wavelength change of scattered X-rays and found "decisive evidence that there is no change of wave-length when X-rays are reflected from a crystal."[66] This must have added to Einstein's eagerness to see whether Compton was right or not.

In two weeks, Mark already sketched an experimental setup and requested Einstein's opinion about it.[67] There is hardly any difference between Einstein and Mark's and Compton's arrangements.[68] Both used a molybdenum target, coal (graphite) scatterer, and an ionization detector.

Einstein often visited Mark, who also lived in Berlin, and his colleagues to discuss Mark's experiments, so there was no need of a vivid correspondence. Furthermore, letters exchanged between Mark and Einstein were lost when the Nazi secret police, the Gestapo, confiscated them from Mark in 1938.

In the end, as Compton convinced his opponents of the correctness of his results, Mark and Einstein considered it superfluous to continue their efforts. At least, this is the way Mark put it in his letter of 1967 to Peter Bergmann.[69] He continued to work on the Compton effect, however, both experimentally and theoretically, together with Kallman, up to 1926.[70]

Einstein again attacked the problem of the nature of light in March 1926 (fig. 2.8).[71] The proposed source of light was again canal rays, as in 1921. As we have seen, according to the wave theory of light, the source of a light wave is the vibration of the electric mass of the atom. According to quantum theory, however, the frequency of the radiation is determined by the energy emitted, with no connection to any vibration of the source.

Suppose that classical electrodynamics holds and that the emitted light propagates in waves, the shape of which follows the vibrations of their source.

Let the radiating atoms, moving in the direction v, be projected on a wire grating by lens G in a way that the images of individual atoms are not larger than the slits between the wires. If the light rays leaving the grating are made parallel by another lens, they will consist of wave trains separated by intervals the size of the slits.

If these wave trains interfere in a Michelson interferometer, interference fringes will appear when peaks meet peaks. This is the case in the figure where

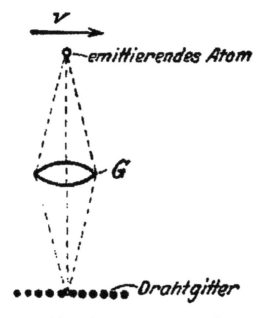

Figure 2.8. The grid experiment. emittierendes
Atom = emitting atom; Drahtgitter = wire grid.
Albert Einstein, "Vorschlag zu einem die Natur des elementaren
Strahlungs-Emissionsprozesses betreffenden Experiment,"
Die Naturwissenschaften 14 (1926): 300–301. Courtesy Albert
Einstein Archives, The Hebrew University of Jerusalem.

the slits are as long as the wave trains. In general, fringes appear if the slit is an even multiple of the length of the wave train. If the slit is an odd multiple, no fringes will appear, for the wave trains cancel each other out. By varying the size of the slits, the interference pattern will change.

Obviously, all this can happen only if light is being emitted over an extended time frame that allows the particles to be projected on several slits. If, however, the light emission takes place in an instant, there are no wave trains to interfere, and somehow the entire wave train passes through a single slit. The absence of interference fringes would prove that the light emission is instantaneous.

Einstein confessed that "I have long since turned the experiment over in my mind." The immediate incentive for the proposal, however, was Emil Rupp's proof of the interference of canal rays.[72] Einstein was impressed by Rupp's paper and by what it demonstrated, namely that the intervals between wave trains could be large enough for his experiment.

Einstein approached Rupp with the request to perform the experiment.[73] He expected to find no interference, that is, that the corpuscular nature of light would be demonstrated. Ehrenfest voted for a positive outcome,[74] and others discredited Rupp's results. Following the well-proven adage "no smoke without fire," Einstein rechecked his calculations and realized that Ehrenfest and his other critics were right. On July 8, 1926, he presented a retraction in a lecture to the Prussian Academy and confessed that "a failure of the classical wave theory . . . seems almost to be excluded."[75] He postponed the publication of this lecture, however, until Rupp confirmed his theoretical prediction, and he presented his paper together with Rupp's results to the Academy in October.[76]

Up till now, I have refrained from putting Rupp's "results" in quotes, but I should not have. As Einstein's correspondence with Rupp demonstrates, Rupp was obliging, even overly eager to "perform" whatever Einstein requested from him. He never lacked explanations why his experiments gave results that did not follow from the setup. He "repeated" the experiments in a day whenever Einstein discovered a flaw in them. More and more arguments were accumulating that contradicted his results—but Einstein stayed undaunted beside him.

In the end, however, Einstein lost his confidence in Rupp's results. Rupp had faked his results, or perhaps had not even performed the experiments at all. In the thirties, he was officially discredited by the German Physical Society, and references to his paper were banned.

As Jeroen van Dongen has observed, Einstein was again misled in using experimental (and "experimental") data by his preliminary expectations, as he had been with Ampère's molecular currents. As soon as he thought that he had got the expected result, he raised no further questions or doubts about their precision or method of evaluation.[77] It was as if he had forgotten his own maxim: a "yes" by nature means only a "perhaps."

All the same, the Einstein-Rupp experiments were illuminating for Max Born, Niels Bohr, and Werner Heisenberg in developing quantum mechanics.[78]

Explaining Superconductivity

When in 1911, in his world-famous and unique cryogenic laboratory at the University of Leyden, Heike Kamerlingh Onnes discovered the superconductivity (zero electric resistance) of mercury at the temperature of liquid helium, 4.2 kelvins, contemporary theories on metallic conduction failed to explain it. Einstein, as special professor at the university, took an active part in finding a satisfactory explanation whenever he paid a visit there between 1919 and 1922.[79]

He proposed a model of "conducting chains." The electrons that constitute the supercurrent move not freely (as molecules in gas) but along definite paths in which atoms serve as stepping-stones. The velocity is determined by the product of their charge e and their (optical) frequency of revolution around the atom v according to Bohr's model (fig. 2.9).

Currents in superconductors consist of an integral number of permanent chains, and the electrons pass along them smoothly. At higher temperatures the chains are torn up by thermal agitation: resistance appears. But let us stay at superconductor temperatures. If a current consists of an integral number of chains and the intensity (the velocity of electrons) is determined by quantum conditions, then there should be no current intensity under a certain limit—a limit that can be measured.

To settle the question, Einstein proposed to put a non-superconducting coil next to a superconducting one and to send a current of growing intensity through the superconducting one.[80] If self-induction in the other coil does not grow parallel with the applied current, then the superconductor does not permit currents under a certain value, and the model of conducting chains is proved. In the optical range, this threshold intensity, ve, amounts to around 0.000015 ampers. The experiment was never performed, perhaps because it was difficult to indicate such a weak intensity.

Another requirement of the chain hypothesis was that the atoms, the stepping-stones on which electrons move as orbiting electrons, should be atoms of the same element; otherwise, the electrons could not pass smoothly. Einstein even surmised that they would not be able to step at all from an atom to a different atom.

This was put to test by Kamerlingh Onnes, twice. He wanted to see whether the interface between two superconductors—in his case, tin and lead—behaves as a superconductor. He found no resistance.

A third proposal involved the Hall effect. When an electric current is sent through a thin slab and a magnetic field is applied perpendicular to it, the negative charge carriers will be pressed to one edge of the slab by the Lorentz

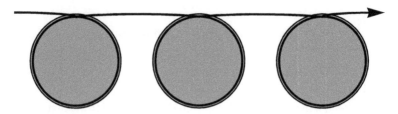

Figure 2.9. Conduction chain.

force, and the positive ones to the other edge. The potential difference between the two edges is the Hall voltage.

From theoretical considerations, Einstein concluded that in a super-conductor slab the Hall voltage should be proportional to its thickness.[81] This conclusion has never been tested, and later, in the light of new discoveries, it lost its importance.

CHAPTER THREE

Expert Opinions

The Patent Office

At the Swiss Patent Office, Einstein's job was to consult with inventors. Only one of his written opinions on patent applications is left, the administrative documents having been routinely destroyed. This opinion is on an alternating-current machine with short-circuit brushes and opposing auxiliary spools for spark prevention. Einstein offers not a single good word for the patent claim: it is "incorrect, inaccurate, and unclear." He also gave the requirements for a correct application: it should note only characteristics of the subject of the patent that are in the claim, and each particular embodiment should correspond to the main claim of the main patent and the claims of the actual patent.[1] His advice did not help the inventor. A revised application was submitted two months later, but Einstein was still not satisfied with it.

We have indirect evidence of a further case. A certain Ignacy Mościcki, inventor of a way to produce nitrogen acid from the atmosphere, of a new method of concentrating nitrogen acid and sulfuric acid, and of high power capacitors, submitted an application to the Swiss Patent Office in 1906. It dealt with an arc furnace for the production of nitric acid, in which the arc was rotated by an electromagnet.[3] As the designated expert, Einstein was especially interested in why the electric arc changed its orientation in a magnetic field. He gave a positive opinion of the application. The story, first aired in 1934,[4] was rediscovered by Zofia Gołąb-Meyer.[5]

That Mościcki and Einstein did meet in Bern is confirmed by an exchange of letters in 1932. Einstein asked Mościcki for help in getting a position for an acquaintance of his. In the introductory sentences, he remarked that "I know that you were originally a physical chemist, and I hear that even now you work as an organizer for scientific and technological research."[6] The letter was addressed "To Mr. President of Poland Professor Dr. Mościcki." Mościcki, the successful inventor and scientist, returned to Poland in 1912, where he was named professor of chemical physics and technical electrochemistry at the Technical University of Lwów (present-day Lviv, Ukraine), and in 1926 was elected president of Poland.

In reply, Mościcki emphasized the meager prospects in Poland for Einstein's protégé, but he also mentioned that Einstein's letter gave him particular pleasure, because it reminded him of their meeting in Bern and later in Fribourg.[7]

From Einstein's next letter we learn that he remembered their meeting with great satisfaction, especially the one in Fribourg with Kowalski.[8] The encounter in Bern may have been in the patent office, but the University of Bern cannot be excluded either, for Einstein succeeded in getting the *venia legendi* (permission to teach) there around February 28, 1908, and he was *Privat-dozent* there from April 21, 1908, to August 4, 1909. In May and June 1908, he also worked at the University of Fribourg in Professor Albert Gockel's laboratory on his first invention, an electrometer for small quantities of electricity (see chapter 4), and on May 24, 1909, he attended a physics colloquium there.[9] In addition, Professor Joseph Kowalski (Józef Wierusz-Kowalski), professor of physics at the university, was interested in this electrometer,[10] so there was ample opportunity to meet Mościcki, for he served as *Assistent* to Kowalski from 1896 to 1912.

Three more patents are supposed to have gone through Einstein's hands:[11] an electrical typewriter with shuttle-type carrier,[12] a gravel sorter,[13] and a meteorological station controlled by ambient humidity.[14]

Gyrocompasses

Hermann Anschütz-Kaempfe, the inventor of the gyrocompass and owner of a gyrocompass factory in Kiel, Germany, sued the American Sperry Gyroscopic

Company.[15] The competition between them had turned fierce when, in 1914, Sperry sold a compass to the German navy. Selling gyrocompasses was a promising business at the time. They could be used on ships, submarines, and airplanes, because metal structure does not distort their indication as it does with magnetic compasses. In addition, in the decade before the First World War, Germany intended to build a navy comparable to or outrivaling Britain's.

One of the Anschütz patents allegedly infringed was DE182855,[16] which protected the original design. Sperry claimed that the patent was void, for it was not new when compared to a previous patent of Marinus G. van den Bos.[17] The other patent that Anschütz claimed to be infringed, DE236200,[18] described a means of damping unwanted oscillations.

The first hearing of November 10, 1914, was adjourned, but the court advised the parties to choose an impartial expert living not far from Berlin in order to keep expenses low. The court submitted a list of potential experts, including Arnold Sommerfeld, professor in Munich, and Felix Klein, professor in Göttingen, who were coauthors of a book on gyroscopes.[19] The first name on the list was, however, Einstein's, perhaps because he was a Berliner, and the court selected him. A court expert's duty was to answer questions impartially, to attend the proceedings, and to provide oral testimony.[20]

The next hearing took place on January 5, 1915. Einstein failed to impress the court; he was not well prepared. To make his task easier, the court formulated four questions to be answered in a written report and presented at the next hearing:

1. What are the physical principles of gyrocompasses?
2. What are the differences between Anschütz's compass and other gyrocompasses, in particular the compass patented by Van den Bos as DE34513?
3. What is the gist of the invention patented in DE236200; and did this invention make it possible to produce the first, perfectly working gyrocompass?
4. What are the similarities and differences between the Anschütz and Sperry compasses, and had Sperry made use of Anschütz's two inventions to such an extent that his compass was technically similar to Anschütz's compass?

Einstein answered the questions on February 6, 1915.[21] He gave a clear exposition of the principles of gyroscopes and gyrocompasses (question 1) and declared that Anschütz was the first to produce the first usable gyrocompass with damped oscillations (question 3), which he achieved in his second patent with a method better than Van den Bos's. Einstein denied the novelty of Anschütz's first patent, giving priority to Van den Bos (question 2). Finally, he declared that Sperry did make use of Anschütz's first patent but doubted that the second was infringed (question 4), that is, he accepted Sperry's damping as an original idea.

Apparently neither the court nor Anschütz was happy with Einstein's voting for Sperry. In the second hearing on March 26, Einstein did not convince the court with his opinion, so he was given two further questions.

The first question practically repeated the previous question 2, asking him to explain the relationship between Anschütz's and Van den Bos's patents. The second question reformulated the earlier question 4: How far had Sperry made use of the ideas in Anschütz's two patents in his own compass delivered to the German navy?

Einstein prepared a supplementary opinion on August 7.[22]

After a meticulous study of the patent specifications, he concluded that Anschütz's first patent did indeed differ from Van den Bos's invention.

To answer the second question, Einstein tested a Sperry compass. He confirmed that Sperry made use of Anschütz's first patent, but, in contrast to his first opinion, he declared that the same is true for the second patent. This time he received no further questions, and the court decided in favor of Anschütz.

Why did Einstein change his mind? Apparently he had not immersed himself in the case deeply enough. When in 1918 Anschütz asked him to serve as his private expert in another proceeding, Einstein expressed the self-critical hope that in the future "the insufficient understanding of impartial experts" would not cause damage to Anschütz and make him angry.[23]

In this case, Anschütz-Kaempfe considered that a patent application of the Gesellschaft für nautische Instrumente (GNI)[24] infringed upon his patent DE241637.[25] The GNI invention was an arrangement to avoid erroneous indication of gyrocompasses when the ship is rolling. He asked Einstein to serve as a private expert on his behalf[26] and solicited an opinion from him on whether the method proposed by the GNI fell within the scope of protection of his patent or not.[27] Einstein's opinion submitted on July 7, 1918, is not available. Apparently Anschütz was not satisfied with it, and he requested another one.[28]

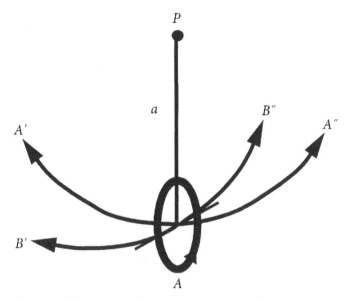

Figure 3.1. Gyroscope motion.

Einstein began his second opinion with an analysis of how the motion of a ship influences the gyroscope (fig. 3.1).[29]

Let the rotating gyroscope be suspended from *P* by a solid rod *a*. If it swings between A'' and A' in a plane to which its axis of rotation stays perpendicular, it will not change its direction. But if it swings *perpendicular* to the previous plane (between B'' and B'), its axis of rotation will oscillate perpendicular to the plane of its swing, but the time average is zero. Consequently, the direction change will not present an insurmountable difficulty in indicating the correct cruising direction.

If, however, the gyro swings in a direction that is a combination of these two directions, a torque may appear that makes the gyroscope's axis of rotation rotate with respect to the vertical. Both the Anschütz and the GNI inventions aim to eliminate this effect.

Anschütz's patent gives two ways to avoid or minimize this effect: to prevent swings from A'' to A' (which is effectively the same as allowing only swings from B'' and B'), or to use two or more gyroscopes with their axes not parallel to each other. If two gyros are mounted in a frame this way, any rotation of the axes is prevented by inertial forces if the angle between the axes is kept fixed by a nonrigid connection. The patent does not stipulate that the gyroscopes be horizontally arranged; the only requirements are that the resulting moment of all the gyros has a horizontal component on which gravitation acts and that the gyro axes be nonrigidly connected and not positioned parallel to each other. Einstein declared that the claims are expounded so clearly that, by following them, any engineer with a knowledge of the subject can build a usable gyrocompass.

Then he turned to the GNI patent application. It also uses two gyroscopes with nonparallel axes and connected with a nonrigid connection, so it is evident that it falls within the main claim of the Anschütz patent. The only question left is whether its construction represents a technical improvement. Einstein said no.

He concluded that the subject of the GNI application fell within the scope of protection of the Anschütz patent, and its specific features are neither novel inventions nor practical developments.

Maybe the applicant succeeded in presenting arguments to prove that these specific features did represent novel invention; maybe the lawsuit took a different course for other reasons. We do not know, because the documents of the court and patent administration are no longer available. We only know that Anschütz lost the case and that a patent,[30] along with two additional patents,[31] was granted to GNI in 1918.

The encounter was, however, not settled at this point. In the spring of 1922, the manager of GNI, Professor Oscar Martienssen, approached Einstein and requested that he withdraw his opinion of 1918 because Anschütz intended to use it against GNI.[32] He argued that Einstein's opinion was based on errors, and he wanted to avoid protracted and time-consuming discussions in various courts that would take Einstein's "invaluable talent" away from "more important things." If Einstein agreed, there would be no need of attacking him in court. From the formulation, one may infer that Anschütz had lodged another suit against GNI.

Martienssen called Einstein's attention to a mistake that he himself had made in an earlier publication,[33] but later corrected.[34] He also added that Richard Grammel followed the same erroneous considerations in his book on gyroscopes,[35] but later he also realized it was a mistake. Apparently Martienssen supposed that Einstein had relied on these publications when he prepared his opinion.

Martienssen's second objection was that Einstein had given no reason why the patent application was dependent on Anschütz's patent. The Anschütz patent states explicitly that the invention refers to a supplementary gyro in the moving system and not to a system with a gyro for stabilizing the cardanic suspension. The GNI application, however, uses a stabilizing gyro. Furthermore, Anschütz's patent can be constructed using two auxiliary gyros but *not* with only one, with its axis perpendicular to the northward directing gyro. The spring that joins the two gyros is of fundamental importance for the Anschütz patent, whereas for the GNI application the vertical gyro is firmly mounted on the ground plate.

"You may rest assured," Martienssen added, "that I feel terrible about writing these lines to you whom I sincerely hold in high esteem for your eminent achievements."

Einstein did not understand the letter, simply because he did not remember the proceedings that had taken place four years earlier.[36] He asked Martienssen for clarification, forwarded Martienssen's letter to Anschütz the same day,[37] and doubted that Martienssen was right, "despite his arrogant tone." The following day, Martienssen replied to him and attached copies of the relevant patents,[38] as well as a copy of Einstein's 1918 opinion. He also added a further objection: in his opinion, Einstein had explicitly stated that Anschütz's direction-indicating system was equipped with two or more gyroscopes and that in the patent of GNI the same system has only one gyro. Why was it then that in his further considerations Einstein mentioned two gyros in the GNI patent application? In it, the stabilizing gyro has nothing to do with the direction indication.

The hearing began on April 11, 1922, in Kiel.[39] On April 12 Einstein prepared a supplementary opinion, but it is not available. Anschütz won the case.[40]

GNI submitted an appeal to the Higher Regional Court on June 9.

Einstein prepared a second supplementary opinion.[41] In the particulars of the appeal, he wrote, a patent was mentioned to demonstrate that there were patents on gyrocompasses with constructions similar to Anschütz's invention but had not been considered to infringe upon it.[42] This patent did not include means to eliminate or minimize the rolling error. The opposite statements in the particulars are untrue.

Then he reflected on the opinions of two experts. The first maintained that the GNI application had a specific feature that Anschütz's patent does not. Einstein replied that the same feature was explicitly described in it and that, at present, it is not the complete independence of the patents at stake but the question of whether the GNI patent is dependent on Anschütz's patent or not.

He summarized his reply to the opinion of the second expert in three points:

1. Anschütz's patent is the first to realize that the rolling error depends on the swing period of the directional system around the gyro axis, and it is the first to offer a means to avoid or significantly reduce this error.
2. The inspected instruments (apparently each firm presented one of its own) use gyroscopes to reduce swinging.
3. In these instruments, the gyros that slow down the swing are not mounted directly on the base of the directional gyro but on a component part, which is connected to the others with rigid connection.

Anschütz's patent would be already infringed if only the first point were a feature of the GNI patent, but because all three points are shared by it, the GNI patent is technically the same as the Anschütz patent.

There was a second session on July 10, 1922. Einstein's main role was as a "bogeyman."[43] Anschütz won again. The last information on this suit is that "the scoundrel did not get away with his tricks"[44]—an impolite and unjust remark from Einstein's pen for, as we saw, Martienssen was polite and raised clear technical objections to Einstein's opinion. While Einstein's first reaction to Martienssen's approach was to call his tone "arrogant," we have ample evidence for Schell's remark, that "the impartial expert had long since become a good and thoroughly partial friend of Anschütz-Kaempfe and his firm."[45]

In Einstein's next case, the defendant was Franz Drexler, and again the plaintiff Anschütz-Kaempfe.[46] Drexler, a trained pilot, had been working on a gyrocompass with Anschütz and, after having left the company, tried to sell it through his newly founded company, the Kreiselbau Co. The invention was a gyrocompass for indication of vertical and horizontal turns of airplanes.[47]

In his opinion of July 23, 1919, Einstein set out by explaining the behavior of a gyroscope that has two degrees of freedom (fig. 3.2).[48]

Let a gyroscope be mounted in an inner gimbal R, which can turn around axis B–B in an outer gimbal G. Springs F restore the position of R whenever it is forced to leave the plane of G. Let the gyroscope be mounted on an airplane with the plane of G parallel to the wings. When the airplane gains or loses

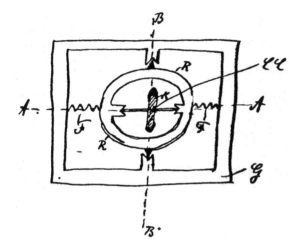

Figure 3.2. Gyrocompass.
Albert Einstein, "Court Expert
Opinion in the Matter of
Anschütz & Co. vs. Kreiselbau
Co.," July 23, 1919. Courtesy
Albert Einstein Archives, The
Hebrew University of Jerusalem.

height, *R* will not leave the plane of *G*, because its axis of rotation *A–A* is shifted parallel. When, however, the airplane turns to the right or left, *R* will step out of this plane by turning around *B–B* and will keep the new position until the turn is finished. Einstein calls this arrangement the "turn indicator."

How can up and down turns be indicated? Maybe with a simple plumb line. Let us suspend such a line *L* in a plane *P* and keep *P* horizontal (fig. 3.3). Until they move uniformly, *L* and *P* will make a rectangle. If *P* is accelerated, *L* will lag behind, and so they will not stay perpendicular to each other. The same happens should *P* turn up or down.

This simple pendulum cannot provide a precise indication of up and down turns, for it is possible that if the airplane is accelerated and turned downward at the same time, *L* and *P* could remain perpendicular. Let us call this device, as Einstein did, the "plumb indicator."

Now return to the gyrocompass and fix *R* on *G* in a line lying higher than the center of mass of *R* and *K*. In addition, remove the springs *F*. By doing so, we combine a gyroscope and a plumb indicator, which, however, cannot indicate directly whether the airplane is turning or not. Einstein calls this a "gyro pendulum."

After these preparatory considerations, Einstein answered five questions put by the court. Because we do not have the original court document, and Einstein's answers do not follow the numbering of the questions, I can only summarize them.

Figure 3.3. Turn indicator.
Albert Einstein, "Court Expert Opinion in the Matter of Anschütz & Co. vs. Kreiselbau Co.," July 23, 1919. Courtesy Albert Einstein Archives, The Hebrew University of Jerusalem.

Anschütz's patent consists of a gyro pendulum and a plumb indicator.[49] The indications of the two instruments must be compared to follow the flight on a general curve. Even though this is not the first patent that indicates the change of orientation,[50] it is the first to indicate vertical turns.

Anschütz can claim priority not in the application of a gyro pendulum but in the combination of this pendulum with the plumb indicator, and, because it is only this combination that can help flight in a curve, the patent represents technical progress of inventive importance.

Drexler's patent follows the same principle as Anschütz's, Einstein continued: it makes use of two gyroscopes with horizontal axes (even though arranged as a turn indicator) and of a plumb indicator. Whether the plumb indicator is a separate device, as with Anschütz's, or it is the turn indicator itself that is turned into a plumb indicator as with Drexler's, makes no difference of principle. There is, however, a technical difference between them: Drexler's device indicates right and left turns directly, not as a difference between the indications of two instruments.

Einstein felt that this opinion sounded obscure for the lawyers, so he appended an explanation, as he put it, in a "freer form," an application of set theory.

Consider a plane P whose points represent all the possible technical realizations of all the patent inventions. The embodiments of a particular invention make a region G of this plane. Had the inventor a complete knowledge of all the embodiments of his invention, that is, of G, he should be considered as the only owner of them. He has, however, only a limited knowledge of G; there can be various technical embodiments of his idea of which he does not know, and which may have novel technical features. Such cases can be called "dependent inventions." Whether such an invention may have some legal rights is a problem for lawyers.

Drexler's patent is a "dependent invention" in this sense: dependent on Anschütz's patent, for it serves the same goal and uses the same gyroscopes, but it is a genuine invention, too, for it uses only one indicator in place of two, and so it produces its result more safely and with greater precision. Drexler's patent covers Anschütz's patent, but it does not copy or circumvent it.

Neither parties were satisfied with Einstein's opinion. According to Anschütz's patent lawyer, Hugo Licht, it was "not negligibly weak";[51] therefore, Einstein was requested to find new arguments in Anschütz's favor.

Einstein called the director of regional court and explained that when he had formulated his opinion, he was staying in Switzerland and, upon returning to Berlin, he found documents that he had not been able to use. He would like to complete his opinion by taking them into consideration.[52] He submitted a supplementary opinion on October 9, 1919, of which only one paragraph is extant, quoted in Licht's letter to Anschütz.[53] In it, he declares that it was Anschütz's patent that first proposed a gyroscope with two degrees of freedom and with a horizontal axis of rotation to keep track of the direction change of the aircraft, and Drexler's turn indicator is also based on this idea.

On the hearing of November 4, under the pressing questions of Drexler, Einstein admitted that if a pilot had known of a paper published in 1910 on a turn indicator,[54] it would not have been necessary to wait for an invention. With this, he weakened the technical importance and priority of Anschütz's 1917 patent. Licht explained Einstein's point by saying that an eminent scientist uses stronger criteria than a judge for what can be considered an invention.[55]

Anschütz won the case, but Drexler appealed. Einstein was again proposed as an expert, but because the court expected a second expert who was "well informed in both the theory of gyroscopes and the behavior of airplane during curved flight,"[56] Anschütz proposed two further candidates, Richard Grammel, professor at the Technical University of Stuttgart, and Ludwig Prandtl, professor of aerodynamics at the University of Göttingen, an international authority and pioneer in the theory of flight. Prandtl confessed that he could not qualify as a practical expert because he had never actually flown, but he could not resist to add, "When Professor Einstein, who certainly has a weaker knowledge of flight than I do, appears as an expert, that will be very interesting indeed."[57] In the end, the court picked a third person, Hans Wolff, an engineer at the German Test Institution for Aviation (Deutsche Versuchsanstalt für Luftfahrt) in Berlin-Adlershof, as a flight expert partner of Einstein. In its session of January 7, 1922, the court solicited a comment on Wolff's opinion from Einstein, which he presented on January 18.[58] Wolff's opinion is not available.

Einstein essentially maintained his earlier opinion that Drexler's invention falls within the area of protection of Anschütz's patent, even though a British patent from 1916,[59] not known to him when preparing his first opinion, might restrict this area. He proposed a formulation of the specific novelty of Anschütz's patent that would stand even in this case as "a clearly arranged combination of a gyro pendulum and an apparatus for indication of the direction of the apparent gravity."

The case ended with a settlement out of court, and Kreiselbau retracted the appeal.[60]

Mixing Tubes

Only the first page of an opinion has been preserved from 1916 on mixing tubes.[61] It is a reply to a plaintiff objecting to Einstein's earlier opinion, namely, that he had not turned attention to the influence of friction on the gas flow in a mixing tube.

To show that he was a knowledgeable person, Einstein assured his partner that he followed "the way of considerations accepted in technical mechanics." He considered a gas flow in a tube and took friction into consideration but neglected gravity. In the special case of the plaintiff's patent, the cross section of the tube is such that the pressure is almost constant along the tube, and this circumstance simplifies theoretical analysis. The page ends with a simple formula for the frictional loss of velocity of unit mass from the inlet cross section to any cross section:

$$\frac{w_1}{w} = \sqrt{1+\zeta}$$

where w_1 is the flow velocity at the inlet of the tube, w its velocity at any of its cross sections, and ζ is a factor of proportionality in the relationship

$$B = \zeta \frac{w^2}{2g}.$$

Here B is the frictional loss, and g is the gravitational acceleration.

Hebeluftschiff

Around the end of 1917, Einstein was "scientific collaborator" of the Mercur Flugzeugbau GmbH.[62] The company's director, Romeo Wankmüller, had earlier been director of the Air Traffic Company (Luftverkehrsgesellschaft m. b. H.) where Einstein's cat's back wing foil was tested around the time (see chapter 4). Apparently Einstein was invited to evaluate the company's new "Aircraft A 28," a *Hebeluftschiff*, a kind of an airship that used compressed air to take off. Einstein's opinion is unknown, but it can be partly reconstructed from a reply to it from Alfred Zehder, Mercur's patent attorney.[63]

Zehder did not share Einstein's skepticism about the efficiency of the machine. Einstein was of the opinion that it is the same whether the compressed air is aimed at an oblique plate from a horizontal nozzle and leaves through an oblique channel, as proposed by a certain Hildebrand, or it runs out directly from an inclined nozzle.

Zehder argued that there *was* a difference and that Hildebrand's idea was better. The reaction force acting on the plates, now called paddles, contributes to the lift. Furthermore, the air streaming along (the oblique channel?) and changing its direction continuously further enhances the lift. There is, however, another important difference between the two arrangements: the air stream pressed against the lower plane (?) will suck the air from between the two planes according to Bernoulli's law, so that each paddle produces a vacuum on the upper plane "as with the aircrafts." I was unable to find a patent on such a peculiar dirigible.

Tungsten Wires for Incandescent Lamps

Sometime around the turn of 1919 to 1920, Konrad Sannig & Co. requested Einstein's expertise in a case brought against it by Osram, a member of the Allgemeine Elektrizitäts-Gesellschaft (AEG, the German equivalent of General Electric). Sannig allegedly had infringed on AEG's patent DE269498[64] with his own patent DE297015.[65] Osram was the producer of incandescent lamps, and the disputed patents described methods for producing tungsten wires for filaments.

Apparently Einstein submitted not an expert opinion for the proceedings but instead an aid to Sannig's lawyers for their preparation for the proceedings, because Einstein concentrated on the formulation of the main claim of the AEG patent and found it inaccurate.[66]

He remarked that before this patent was submitted (October 6, 1910), the only process known for the production of tungsten wires was the mechanical treatment (hammering) of the initial tungsten powder rods into wires at high temperature. The end product was, however, not sufficiently ductile and not suitable for cold wire drawing. AEG was the first to find a process to produce wires without these drawbacks—namely, more hammering than necessary to reduce the initial cross section to its final value.

With the advent of chemically purer and less porous tungsten, however, this additional mechanical treatment lost its significance, and hammering could be reduced to shaping the cross section.

Relying on this brief historical introduction, Einstein criticized a passage in the AEG patent: "hammering repeatedly for a longer time until they turn ductile and can be drawn at usual temperature." He stressed that this claim does not exclude "with sufficient significance" that the required ductility can be achieved in other ways than by repeated hammering. At the end of a short summary, Einstein again rephrased AEG's main claim: "Flexibility and ductility of wires in a cold state is achieved so that during their production these wires are subjected to more mechanical treatment than would be necessary for the mere *forming* of the wire."

Sannig's lawyers read the opinion with some puzzlement and turned to Einstein with the question of whether it can be interpreted as saying that processes that do not make use of mechanical treatment more than necessary for shaping the material are to be distinguished from processes that do use additional mechanical treatment. If they are, their patent, falling in the first category, cannot infringe upon the AEG patent, which falls in the second one.[67] Einstein gave an affirmative answer.[68]

Then, two years later, the lawyers of AEG found his amended formulation of the main patent claim ambiguous and asked whether Einstein considered a process for cold production of ductile and flexible tungsten wire not patentable even when the starting material is completely nonductile and can be made ductile only by mechanical treatment. He denied this interpretation.[69] He, however, added that now he would prefer another formulation: "A process to produce cold ductile tungsten wires for incandescent lamps characterized by the fact that the cold ductility is produced by mechanical treatment of the originally cold-brittle tungsten body." This way nonmechanical treatment for achieving ductility would be excluded. In his original formulation and in his correspondence with Sannig, he had assumed as correct Sannig's statement that, in their process, the mechanical treatment served only the shaping of the material and had nothing to do with ductility.

Triodes for Amplification

In the fall of 1920, Georg Count von Arco, one of the directors of Telefunken, the German company for wireless telegraphy, asked Einstein to serve as an expert in a lawsuit.[70] Arco and Einstein were longtime acquaintances for they were committee members in a pacifist organization, the Bund "Neues Vaterland."

The inventor of the patent allegedly infringed was Alexander Meissner, an engineer at Telefunken,[71] and the infringing patent had been granted to Ludwig Kühn and Erich F. Huth GmbH.[72]

Einstein's opinion is available in a rather sketchy handwritten draft that was prepared sometime after Arco's letter.[73]

The first question he raised regarding Meissner's invention was whether Kühn's invention was a novelty. He established that Meissner's invention, though not the very first to use "ventil-like arrangements" that utilize "electric discharge processes," was the first to present a practical method of how to use "grid vacuum tubes" (triodes) for transmission and reception of radio waves. Kühn's patent does not have a special device to make the coupling of anodic circuit with the grid circuit; it is the capacitance between anode and the grid that takes care of it. Even though Einstein admitted that this difference is something new when compared to Meissner's present and earlier patents,[74] he did not accept it as a novelty—but he did not say why not. The only feature that Kühn's patent may claim as something new is that the arrangement is used not only for reception but also for transmission of radio waves. In the preceding paragraph, however, he attributed this capability to the Meissner arrangement, too. This contradiction can be explained if the manuscript represents an early version of the final opinion. In conclusion, Einstein added that this patent contains nothing that can be considered an invention.

The second question was whether the Kühn patent represented a substantially new *technological* advantage over the Meissner patent. Kühn claimed that it offers precalibration, weaker influence by the defects of the antenna, and weaker harmonic waves (at the time these outer influences were a major obstacle to long-distance telegraphy and radio transmission), but these features are not connected with the specific coupling. They are concomitant with all loosely coupled oscillation circuits.

Einstein's summary opinion was that there is no new technical development in Kühn's patent.

Sound Direction Ranging in Air and Water

At the end of 1921, Einstein prepared an opinion on a case[75] where the invention of the plaintiff, Signal Co., was a device to determine the direction of a sound source in water.[76] The invention proposes arranging microphones at relatively distant intervals along the hull of a ship to indicate the direction of a signal's source (e.g., an underwater bell) by the difference between the times that the signal arrives at the microphones (fig. 3.4).

Einstein first described what he considered to be the novelty in the plaintiff's invention. He compared it with a German, a British, and an American patent, patents that may have been proposed by the court for comparison. The German patent uses a different method, not based on the time difference,[77] so it is irrelevant. The sound detecting apparatus of the British and American patents[78] is the human ear, and in contrast to this subjective method, the plaintiff's patent uses an objective method for detecting and measuring the time difference (namely, instrumental measurement of difference in arrival times), and it is the first to offer such a method. The larger distance between the microphones improves precision, even though the apparatus is more complicated. None of these ideas can, however, be considered original.

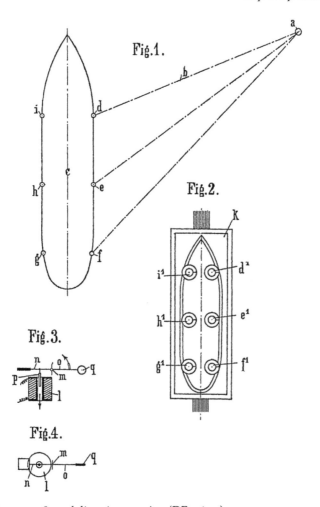

Figure 3.4. Sound direction ranging (DE256747).

Then he turned to the patent of the defendant, Atlas Works. The patent is not mentioned by its number. From a second expert opinion in a case between the same sides, it is highly probable that it is DE301669,[79] a stand with two microphones on a horizontal rod about two yards from each other and with tubes leading to the ears of an observer. The microphones are turned around with the stand until the observer hears the sound coming from the direction perpendicular to the center of the stand: that is the sought-for direction (zero method). The inventors of the patent, Erich M. Hornbostel and Max

Wertheimer, worked at the Psychological Institute of the University of Berlin, and Wertheimer was acquaintance of Einstein's.[80]

This patent also uses microphones, but the time difference is measured not with a device but with the human ear. It differs from the British invention in that it uses a zero method, which results in higher precision. In contrast to the plaintiff's patent, here the microphones need not be mounted "at a larger distance," which, as Einstein calculated, must be at least 50 meters (57 yards) to detect 0.01 seconds time difference. The defendant needs only a 0.9–1.8 meter (3–6 feet) distance. All in all, the features common to both inventions had been known earlier, and the devices for objective indication of time difference added to the plaintiff's main patent in his additional patents are missing in the defendant's invention. The subjective method is preferable because the human ear was 10 to 100 times more sensitive than contemporary instruments.

Einstein concluded that "the defendant's apparatus does not make use of ideas and supplies offered by the plaintiff's patent, the novelty of which had not been anticipated by the British patent No. 15102 (and American patent No. 224199)."

On December 3, Einstein prepared another opinion for the same sides. However, while according to Einstein's earlier opinion, the first case was "Signal versus Atlas," in the title of this second opinion, the case was "Atlas versus Signal." The patent DE301669, the patent of the defendant in the previous case, is now the plaintiff's patent, so we may assume that the second case was a countersuit.[81]

Einstein's task was to evaluate the novelty of the plaintiff's patent.[82] He did so by answering questions put by the court.

The patents mentioned by the defendant as anticipating the plaintiff's patent[83] have nothing to do with it, Einstein began. An American and a French patent[84] do use the human ear to find the sound direction as the plaintiff's patent does, but the novelty of this patent consists in the possibility of changing the length of the pipes leading from the microphones to the observer's ears. There is no need to turn or adjust the microphones to hear the sound from the central direction (a difficult operation when they are mounted on the hull of a ship or are placed at a significant distance from each other). This is "a novelty of inventive significance." The defendant does not make use of means used in the plaintiff's patent, Einstein concluded.

Prospecting for Ore and Water from a Dirigible

Heinrich Löwy, a Viennese inventor, had long been interested in prospecting for mineral resources from the air. It had already been known earlier that electrically conducting materials such as water and ore can be searched for by an electrodynamic method. Prospectors trailed an antenna on the ground, induced high-frequency electromagnetic waves in it, and mapped the change of its capacity with the materials in the deeper layers of the soil. Löwy's innovation was to trail the antenna not by hand but by a dirigible, thereby extending the applicability of the method to large arid areas such as deserts or, when flying at a height of 150–300 feet (the usual altitude of aircraft at the time), even to forested terrain. He published papers and a book on this method of exploration,[85] and test flights had also been conducted in Germany and its colonies.

The method promised great success and numerous benefits, especially after the First World War, when military dirigibles were being turned into commercial carriers. A syndicate was founded with Löwy as manager, in cooperation with Theodor von Kármán, professor at the University of Aachen, and Richard von Mises, professor at the University of Berlin.

In summer 1921, Löwy forecasted a boom in long-haul commercial dirigible flights, for example, between London and India, or New York and San Francisco, and saw in his method, when used over the huge arid territories of the United States and the Middle East, a means to make commercial flights economic. He proposed to raise the capital of the syndicate and to seek the involvement of the eminent German dirigible-building company, Zeppelin Works. He launched a "propaganda campaign" and turned to four specialists for opinions on his method: to Einstein as the physicist, to Richard von Mises as the flight specialist, to Theodor Liebisch as the geologist (all professors of the University of Berlin), and to K. W. Wagner, director of the Imperial Office of Telegraph Technology (Telegraphentechnische Reichsamt) in Berlin as the electrotechnical expert. They all came up with glowing opinions.[86]

Löwy announced that he had applied for a new patent in Germany,[87] "which is due to Einstein's and von Mises's propositions and, in my view, may represent the final solution of the task."[88]

In his short opinion,[89] Einstein proposed a technical test of Löwy's proposal but nothing else. Further details can be found in the letter in which he was informed that 5,000 marks had been transferred for his services[90]: "Your proposal to lift the antenna from the ground completely and to compensate for the increase in height by using more sensitive measuring methods is of great practical importance for us," the letter remarked. The same proposal can be found in Mises's opinion and acknowledged by Löwy as due to Mises.[91] We do not have enough documents to decide whose proposal this idea actually was. The available sources are extant only because Löwy sent copies to Theodor von Kármán, his consultant and friend, and were preserved in Kármán's estate.

Riveting Hammers and Pile Drivers

From 1921 Einstein served as a consultant on inventions to Rudolf Goldschmidt for an annual remuneration of 18,000 marks,[92] half of his annual salary from the Prussian Academy at the time. Goldschmidt was the head of a research laboratory or workshop, located at the Bergmann-Elektrizitäts-Werke, Berlin.[93] Whether it was part of the company or privately owned by Goldschmidt is unknown. His most famous achievement was the high-frequency radiotelegraph inaugurated in Berlin on June 19, 1914, a month before the outbreak of the First World War, with an exchange between Kaiser Wilhelm II and President Woodrow Wilson.

We can only guess what Einstein's consultation consisted of. Before the date of the letter, Goldschmidt had submitted four inventions to the German Patent Office, all on transformation of rotary motion into reciprocating motion with the goal of constructing electric hand hammers. Actually they were submitted by Det Tekniske Forsøgsaktieselskab, Charlottenlund, Denmark, between November 10, 1920, and February 7, 1921, and were accepted as German patents.[94] (The patents do not mention Goldschmidt's name, but an American patent, submitted on January 10, 1921, and accepted as US1386329,[95] closely follows them, and names Goldschmidt as inventor and assignor to the Danish company.)

In January 1922 Goldschmidt asked Einstein an additional favor: to assist him with an expert opinion in a patent case.[96] As we learn from Einstein's reply, his problem was whether any of the fourteen American patents Goldschmidt had sent to him lessened the importance of his own American patent (US1386329).[97]

Goldschmidt's patent refers to "a mechanism for converting rotary into reciprocatory motion," but the invention offers more: an automatic movement of the mechanism in the direction of the reciprocatory motion. As an example, the patent description mentions hand tools (e.g., a chisel, a riveting hammer), forging hammers, and pile drivers.

Among the fourteen patents that Einstein glanced through, he found a concrete tamper, a massage device, a car horn, and shaking mechanisms. His summary statement notes that even though all of them are on a transformation of rotary into reciprocatory motion, none of them adds a means of translatory motion, and none of them limits the importance of Goldschmidt's patent.[98]

Production of High-Pressure Gases

Early in January 1922, Paul Hausmeister requested Einstein's opinion on his invention.[99] Hausmeister carried out electrolysis of water in a closed container. The developing hydrogen and oxygen had a pressure almost 2,000 times higher than the atmospheric pressure. He found that the electrical work necessary for the electrolysis did not depend on the pressure of the developing gas. He concluded that there is no need to let the gases dilate to reach atmospheric pressure and then compress them to fill in gas cylinders.

Einstein was impressed by the invention, even offering a formulation for a patent claim: "A process for the production of high-pressure gases, namely, liquids are decomposed electrolytically under high pressure with the goal of dispensing with a compressor for the gaseous products of electrolysis."[100]

Hausmeister applied for a patent on the production of gases under pressure at the end of 1923.[101] The phenomenon discussed with Einstein is, however, only mentioned cursorily. In their later correspondence, no patents come up.

"Electrophonic Piano"

This story goes back to 1912, when Richard Eisenmann, a Berlin citizen, submitted an invention to the British Patent Office and a year later to the Austrian Patent Office on a rotary interruptor of electric current to produce musical tones.[102]

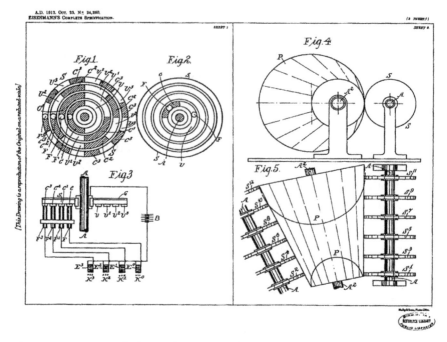

Figure 3.5. Eisenmann's electrophonic piano (GB24260 A. D. 1912). *Fig. 1* and *Fig. 2*: discs with conducting and isolating stripes. *Fig. 3*: top view of a disc, with axis of rotation *A*. *Fs* are contacts closing and opening the electric circuit consisting of electric source *B* and electromagnets *E* moving strings *K*. *Fig. 4* and *Fig. 5*: Cone *P* rotating discs *S*.

The basic idea is to make wires vibrate not by hitting, dragging by a bow, or strumming them but by using electromagnets for pulling and releasing them at the eigenfrequency of the wire or its multiples. Revolving discs with conducting and insulating sections serve as interruptors, and the twelve discs necessary for covering an octave are driven by a common cylinder or, as in figure 3.5, by a cone. If we shift a disc along its axis, its speed of revolution, and with it the frequency of interruptions, can be set precisely.

Eisenmann's goal was "to provide a mechanism that enables one to make the piano's tones sound any long and to swell them."[103]

In 1914 Eisenmann came up with another invention. As he wrote in the patent description, "According to the invention, a brake is associated with the motor shaft, and controlled electro-magnetically through the agency of suitable clockwork, which is utilized to constantly impart to the movable element of a controller the same number of rotations as that which the motor is to make, the arrangement being such that the braking action is automatically increased when the motor commences running too rapidly, and is decreased when its speed is too low."[104]

So far no mention of Einstein. In July 1922, however, Eisenmann invited him to test the piano. His letter of invitation is missing, but he expressed his thanks for Einstein's visit on July 18. More than that, he reminded Einstein of his promise to give "a statement and an opinion." He added a technical description of the device to make Einstein's task easier.[105] And now we learn that the rotation of the motor driving the discs was kept uniform by making use of this second patent, namely a "clockwork," a pendulum. If the motor rotates the cone and, through it, the interruptor discs irregularly, the vibration of strings will not be constant and the tones sharp. This special piano "adds to the cold sound of the piano the richness and volume of an organlike but metallic sound of an orchestra, as had been desired by Franz von Liszt," Eisenmann wrote. I am sure he had in mind Franz Liszt, the virtuoso pianist, and not Franz von Liszt, his cousin, who was professor of law at the University of Berlin.

We also learn what the purpose of Einstein's invitation might have been: "Now the work on it has reached the stage when a person with capital and energy can make use of it on an industrial scale." Apparently he hoped that Einstein's name would help find such a person.

Einstein kept his promise and gave "a statement and an opinion" in three points: the tone of the device is of a quality that will add to the means of musical expressivity; the technical details are so simple that any skilled expert can build it; and the control of the uniform operation of the motor is unique. This third characteristic Einstein called "your main achievement," and he was of the opinion that this method can be used in the most diverse fields of precision measurement.[106]

I do not know whether this control method has been used in precision measurements, but Einstein's opinion that the novel instrument would augment the possibilities of musical expression has stood the test of time. With Eisenmann as a pioneer, the so-called electromagnetically prepared piano has been picked up by composers of our day, as we can learn from a paper of one of them, Per Bloland of Stanford University, who not only gives details of the up-to-date form of the instrument but also analyzes its compositional possibilities.[107]

Aerial Stereophotography

"First of all, a reminder of our time spent in the Luitpold Gymnasium. You lived then on the Rengerweg in the Alhambra and I at the corner of Lindwurnstrasse. There I saw the first arc lamp in front of your father's factory." These are the introductory sentences of a letter, written in Munich in 1948 by Max Gasser.[108] He reminded Einstein of a patent case in 1922 in which Einstein gave a favorable opinion of his invention, a process and device for preparing stereographic maps from aerial photos. He complained about the obstacles the big German optical firms such as Zeiss and the military geodesic service put in his way. He even approached the American Geological Survey, the Army Corps of Engineers, and other organizations, and they all spoke favorably of the invention. "In the USA technical literature, my Aeromultiplex sails under foreign colors, that of Zeiss. Therefore the surveying companies have heard nothing of the invention nor of your prescient opinion."

It was not in 1922 but in 1923 that Einstein had been invited to serve as an expert in the proceedings of Internationale Aerogeodätische Gesellschaft, Berlin (Inag) versus Optikon GmbH in Dresden, a company founded by Oskar Messter, a pioneer inventor and producer of movie projectors and movies. The invited experts were supposed to inspect the competing instruments in Berlin and Dresden, respectively.[109] Einstein, however, gave his opinion before these inspections, on April 4, 1923.[110] Because he called Gasser's patent "plaintive," it is highly probable that it was Inag that represented Gasser. If we add that, in his letter to Einstein, Gasser mentioned a company that he had founded to circumvent obstacles that big companies put in his way, we may identify Inag as Gasser's company. The Optikon patent that infringed Gasser's is not mentioned anywhere; it may have been any of Messter's patents on aerogeodetical procedures and equipment.[111]

Going into detail, Einstein assured the lawyers that the evaluation of the invention is not as complicated for their legal purposes as it looks from its technological details.

In Einstein's opinion, the novelty of Gasser's invention consists in the combination of already well-known methods:[112] methods of how the position and orientation of a photo camera can be reconstructed from a relief picture when the actual positions of three points of the relief are known, and methods and devices to reconstruct the form of the object by central projection of singular points of two photographic pictures, when these are brought in the same relative positions where they had been taken.

Furthermore, with his additional patent,[113] Gasser was the first to produce an apparatus for the implementation of the procedure and made it possible for a single person to see the pictures of corresponding points at the same time.

Gasser's invention is a pioneer patent, Einstein concluded, and there is no doubt that the defendant's apparatus falls in the domain of protection of Gasser's patent.

With this opinion, Einstein's role in the proceedings came to an end.

Magnetic Cores with Low Electric Conductivity

In 1928 Einstein was asked by Siemens & Halske AG to prepare an expert opinion for Deutsche Kabelwerke, a company close to Siemens & Halske.[114] Kabelwerke sued Standard Telephones & Cables, Ltd., for patent infringement. In his opinion, Einstein repeatedly insisted on the priority of the patent DE341678,[115] indicating that this must have been the patent owned by Kabelwerke and the other, DE390178, the Standard patent allegedly infringing upon it.[116]

Einstein began by giving a short overview of the earlier similar inventions. The first patent on how to produce magnetic cores to avoid induction currents in them was issued in 1884.[117] It proposed mixing finely cut pieces or powder of magnetic metal with resin or shellac and compressing the mixture under heat. Fiber or hair can be added to strengthen their cohesion.

Cores produced by using hardening binder may have low electric conductivity and satisfactory mechanical resistance, but they cannot have the magnetic permeability that thin iron cores have because, even when we neglect the isolating binder layer, two powder particles touch each other over a practically infinitely small surface.

The patents in the litigation offer methods to avoid this drawback. DE341678 proposes to press the particles beyond the limit of elasticity; DE390178 claims the same but with a different means to produce the isolating layer between particles. The high pressure flattens the particles at their point of contact and lowers their magnetic transmission resistance. In addition, the particles are bound together not with a binder but with a pressure of 10,000 to 15,000 atmospheres. The method was first offered by patent DE341678.

Efforts had been made earlier to produce cores with high pressure but without specifying the pressure.[118]

That the flattening effect and the binding effect can simultaneously be achieved by pressure is plausible, retrospectively, because to pass the limit of elasticity means a relocation of microcrystals, which is a precondition for the cohesion of the crystals. The existence of this double effect of such a pressure was first mentioned in the application for DE341678.

Einstein also criticized the opinions of two other experts. They did not mention the importance of flattening in achieving satisfactory permeability, nor did they give weight to the fact that pressures previously uncommon in electric industry must be used.

Telescope for Daylight Observations of Phenomena near the Sun

On March 13, 1928, a patent application was registered for Ivan N. Kechedzhan in the Soviet Patent Office on a telescope for observation of phenomena apparently near the sun.[119] In the patent documentation, his Russicized name Kechedzhiev was used. His goal was to present a means of testing one of the consequences of the theory of general relativity: the deflection of light by gravitation.

This had already been confirmed in 1919. British astronomical expeditions observed that the sun deflected the light rays emitted by stars apparently near its disk, but they had to seize the opportunity of a total eclipse, when the sun's overwhelming brightness is covered by the moon. Kechedzhan offered a telescope for the same observation in full sunlight (devices for observations near the sun were later called coronagraphs). He proposed a metal frame of square cross section and 35 meters (115 feet) long with its inside painted black. A metal disk of the visible size of the moon would be mounted at its upper end that could be moved on a pole to shield the sun. A dark chamber with a small telescope would be attached to the lower end of the frame.

In the fall of 1929, Kechedzhan proposed to the All-Union Society for Cultural Relations with Foreign Countries that an expert opinion be requested from Einstein. The society forwarded the request and the patent description to Einstein on February 19, 1930. (In it, Kechedzhan is mentioned as Katschedjan.)[120]

Einstein answered in a week.[121] The long frame for exclusion of optical disturbances of the scattered solar light is already well known, he wrote, but the disk to cover the solar disk and exclude its direct, intensive light is useless. To make it even somewhat effective, it would have to be mounted at an extraordinarily large distance from the objective of the telescope, namely at its focal plane. This is also well known to specialists. To sum it up, Kechedzhan's proposal is worthless.

Makeup Mirror

This time it was a member of his wider family, cousin Walter Kocherthaler, who in 1934 turned to Einstein for advice on whether his patents had been infringed or not.[122] He was convinced that the task would be child's play for Einstein, because he remembered the halcyon days in his home in Berlin-Dahlem where Einstein, fascinated by the effects the sunlight produced through the closed shades of the guest room, tried to express them in mathematical formulae, daubing "long lines of hieroglyphs on toilet paper."

Kocherthaler, his business partner Peter Schlumbohm, and the French glass factory Saint Gobain managed to produce a special sort of glass. The mirror made of this glass was meant for ladies who wanted to check their makeup. If they looked in its golden-colored side, they saw themselves as if illuminated by artificial light; if they used the other, blue side, they looked as they would in daylight. Its brand name was Pre-Vue Mirror.

The invention had been patented in the United States, and now Kocherthaler was on the way to producing the mirrors in commercial quantities and selling them to leading cosmetic concerns. Not long before, however, another company entered the market with a lower-quality imitation. Kocherthaler intended to launch a suit against it and to turn to the National Better Business Bureau with a complaint about lower-quality products. He wanted to reinforce the case with a scientific explanation, and this was what he requested from Einstein. He attached copies of patents and two mirrors. They are not available.

In his reply,[123] Einstein mentioned only one patent by number, US1951214.[124] He started by saying that he looked at the case as a patent expert and not as a physicist. If at the time of application it had been considered a novelty to produce a makeup mirror, which, because of its considerable absorption of short-wave light, was capable of and meant for showing a person in daylight as if she were in artificial light, then any mirror capable of doing the same and meant for the same purpose falls within the domain of protection of the patent. If, in addition, the other invention also uses colored glass, no doubt it infringes the Pre-Vue patent. Finally, if the same goal is achieved by a double mirror that is capable of producing the opposite effect, too, and is put on the market, it is an even more serious infringement of patent law, if one presumes that such double mirrors had not yet been known at the time of application for the Pre-Vue Mirror.

Kocherthaler was not very happy with this opinion.[125] Politely blaming himself for the "not quite correct formulation" of his request, he explained that his problem was whether the statement that the mirror reflects images in daylight that look as if they are illuminated by artificial light and vice versa is correct exactly or only approximately. And, if only approximately, he wanted to know whether this divergence is small enough to consider this glass as "the closest to a mathematical ideal in future practical applications."

Einstein's answer is unknown.

Balanced Tapered Bearing Rollers

On August 7, 1944, Einstein received a letter from Otto Henselman, an American inventor in Mount Eaton, Ohio, asking him to explain why balanced tapered rollers do not heat up under high speed as soon and as much as unbalanced ones.[126] To explain what a balanced roller is, he attached his patent description (fig. 3.6).[127]

The roller rolls around a cone in such a way that its wider end rolls on a circle larger than that on which its narrower end rolls. As a consequence, the wider end, heavier than the narrower end, is subjected to a higher centrifugal force. To eliminate this difference, Henselman proposed forming a cavity in it, large enough to make the centrifugal forces equal at both ends. This, he claimed, will make the roller more efficient and durable.

I am tempted to add another reason why Henselman sent Einstein this patent description: he wanted to impress Einstein with his "knowledge" of science. His *Fig. 5* shows "the centrifugalised area in which the . . . rollers operate." They are simply points where the roller is subjected to highest pressures (A and B) and no pressure (C and D) when the bearing is supposed to be under pressure from above (e.g., in the wheel of a vehicle). "The discovery of these four points . . . as points of event, relative to time, and space, and the law of the

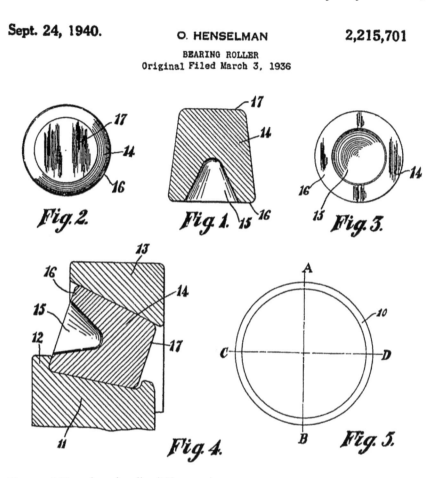

Figure 3.6. Henselman's roller (US2215701).

force of the gravity, we are inclined to accept as 'true.'" Do not ask me what this collection of terms means. The patent description offers other gems of obscure formulation, which, with some effort, can be translated into understandable technical terms.

Henselman does not seem to be a trained engineer or physicist. Spelling errors in his handwritten letter ("set fourth," "inclosed," "in deed," "Copie" always capitalized) suggest maybe he was a first-generation German. This guess, however, is contradicted by his consistently calling Einstein "Enstein."

Anyway, Einstein answered Henselman's letter in a few weeks.[128] He was of the opinion that the improved durability of the roller was the consequence of its higher deformation due to the cavity: the larger surface of contact between roller and bearing lowers the pressure on the spot of contact. He even proposed replacing the cavity with a hole to extend this deformation to the entire roller. "This explanation is only a guess," he added, and apologized for the late answer: "I am overburdened with correspondence." No doubt he was, for in his haste he did not realize that he did not answer the only question that Henselman asked: why balanced and unbalanced rollers heated up differently.

CHAPTER FOUR

European Inventions

The "Little Machine" (*Maschinchen*)

In April 1908 Einstein published a short paper on an electrostatic method for the measurement of small quantities of electricity.[1] He was led to the idea by the following considerations.

In his papers on Brownian motion published between 1905 and 1908,[2] Einstein showed that the observable motion of very small particles suspended in a fluid (in the observations made by Robert Brown, these were pollen) can be explained by the thermodynamic characteristics of the fluid. The fluid must consist of molecules in thermal motion that collide with the particles. This was a strong argument for the existence of molecules or atoms, a view not shared at the time even by such prominent physicists as Wilhelm Ostwald or Ernst Mach.

At the end of 1906, the idea struck Einstein that voltage fluctuations must occur in condensers because of the thermal agitation of the molecules that make up the condenser.[3] These fluctuations, analogous to the fluctuations in a fluid, would give rise to random electric charges. If so, then he would find an argument for the atomistic constitution of electricity. In 1907 he already had an idea of how to construct a device for the measurement of this very small potential difference and communicated it to his friends Conrad and Paul Habicht.[4] In a month he was happy to learn that the Habicht brothers had already built such a device.[5] In September he felt a "deadly curiosity" to know what the Habichts were doing.[6] He played with the idea of patenting the invention but gave it up because of "the lack of interest from manufacturers."[7] However, he submitted a manuscript on it to the *Physikalische Zeitschrift*.[8] Paul Habicht was delighted that Einstein "took care of his priority."[9] With not a single word on his own efforts to develop the *Maschinchen*, Habicht spent the major part of his long letter describing his new invention, a flying machine that looks like a kind of helicopter with double horizontal propellers that were to rotate in opposite directions. Einstein realized that Habicht had lost his enthusiasm and approached Adolf Gasser to recruit a mechanic from the local *Technikum* to continue.[10] Their extant correspondence consists of a single letter from Gasser; the other letters, if they existed at all, ended up, as the gossip says, on a nail driven in Einstein's doorpost only to be dropped into the waste bin when the nail was full.

The paper puts the question how to measure small electric charges.[11] Contemporary quadrant electrometers took the energy for the deviation of their needle from the energy of the system to be measured. This was the reason why the sensitivity could not have been increased over 0.000001 volts, Einstein claimed. He proposed to use an auxiliary source in the form of an influence machine to supply the energy for the motion of the needle.

Figure 4.1 is a schematic top view of the instrument. A_1 and A_1' are fixed conductors, which the metal plates B pass by. The small plates can be mounted on a wheel (as with a usual influence machine). The springs K_1 and K_1' contact the fixed contacts b when they meet during the rotation of the wheel. K_1 is put to the earth.

Let us keep A_1 on a positive potential P_1. When the contact b of plate B touches K_1, the electric charge on A_1 induces an opposite charge $-e$ at b.

This charge is carried by B to face A_1' where K_1' takes off the charge and carries it to A_1'. This charge transfer will go on until a stationary condition evolves, with negative potential P_1', which is proportional to but independent of P_1. If the As are substituted by collectors that take off the charge from both sides of B, this factor of proportionality a can be higher than 1, for example, 10. If, in addition, we use a cascade of n identical influence machines, the final potential will be $P_n' = a^n P$. By increasing the number of cascade elements, the sensitivity can be increased, the only limit being external error sources. The

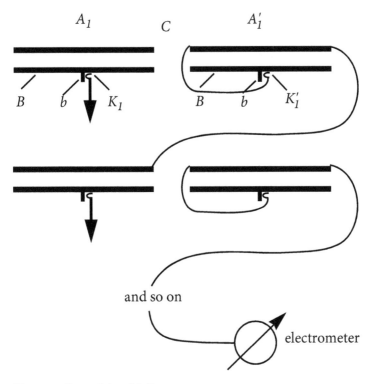

Figure 4.1. Potential multiplier.

energy required for setting the electrometer in motion is no longer the electric energy of the system to be measured but is an external mechanical energy.

At the end of the paper, he called upon physicists to build such a device and offered additional details for those interested.

The first and, as it turned out, the last comer was Joseph Kowalski, the Fribourg professor whom we met in connection with Mościcki's patent. He intended to build the machine.[12] Apparently Einstein supplied further information, because Paul Habicht mentioned in one of his later letters that Kowalski struggled with the contacts and the isolators.[13] Habicht would not have been a Habicht if he had not offered Kowalski, via Einstein, an invention: an alternating current gauge.

On April 4 Habicht gave an enthusiastic account of his old-new idea of a sound amplifier for telephones. As a coup de grace, he remarked, "It is a great pity that no one is set to work on your *Maschinchen*. But maybe you have been in too much of a hurry[;] it will turn out all right."[14]

Upon Einstein's (nonextant) request, Habicht sketched three contacts for the *Maschinchen*, but he could not resist adding two further ideas: a circuit breaker for dynamos and a "tire" made of wood splints, to be used on cars or even on the railroad.[15]

According to the diary notes of Albert Gockel, a colleague of Kowalski in Fribourg, Einstein worked in Gockel's laboratory on June 28, 1908,[16] perhaps giving Kowalski a hand in building his *Maschinchen*.

But Einstein started working at his Bern home, too. With a good mechanic, he built a second *Maschinchen* and an electrometer, to test it for potentials below 0.1 volts. "You wouldn't be able to suppress a smile if you saw the magnificent thing that I patched together myself," he said in a shy remark to Jakob Laub.[17] By December the device could measure potentials less than 0.001 volts.[18] If he could raise the sensitivity to 0.00001 volts, "then nothing stands in the way of an experimental test of the limit of validity of electrostatics required by the molecular theory." He felt satisfied: it was no mean achievement to perform experiments in so short a time, without a laboratory, and with his own money.

Einstein was happy to have the *Maschinchen* back from the mechanic. "I am busy doing experiments; I now have ready to hand everything that I need for investigating the limits of the applicability of this method." He checked the contacts and coherer with a colleague at the patent office.[19]

Einstein did not break off contact with Gockel. Around March 25, 1909, he planned to visit him in Fribourg. The experiments performed in Bern indicated that the main problem was the mercury contacts invented by Paul Habicht, and he intended to test them in Fribourg.[20] In mid-April he borrowed an electroscope for preliminary measurements and announced to Conrad Habicht that the multiplication had been raised over 200,000.[21]

Toward the fall of the year, the development of the device was taken over by the Habicht brothers. A resigned query about its problems ("Where does the poor devil begin?") is all Einstein sent to them in September,[22] but in November he invited the brothers to his home for Christmas to work together.[23] Einstein made a second attempt to meet with them in March 1910[24] to discuss not only the device itself but also the Habichts' paper on it.

The Habichts developed the *Maschinchen* and supplemented it with auxiliary devices. On December 15, 1911, Paul Habicht demonstrated the small device to the German Physical Society in Berlin,[25] and, as Einstein remarked, with it he accomplished a breakthrough for the *Maschinchen*.[26] Einstein thought it would soon replace the contemporary sensitive quadrant and filament electrometers, and while he was pleased by its success ("folks stood breathless," he wrote to Michele Besso),[27] Paul Habicht was pleased by the failures of its competitors.[28] Habicht gave a second demonstration to the Swiss Physical Society in Bern on March 9, 1912.[29]

Even though further improvements were made to the device, disturbing effects were discovered, common to electrostatic devices: atmospheric and contact electricity, as well as friction-induced electricity. In 1927 Habicht founded a firm to produce and sell the electrometer, but only a few were sold before the 1930s. Perhaps the idea was not novel enough: electrometers with the sensitivity requested by Einstein had been available since 1906, and to develop a device that was merely a variant of existing machines is a telltale sign of Einstein's ignorance of contemporary electrometers. The information he had about them apparently came from the courses that he had attended at the Poly in the winter semester of 1897–98.[30]

Planimeter

On February 28, 1914, Einstein delivered a lecture to the Swiss Physical Society in Basel on a method for the statistical use of apparently irregular, quasiperiodic observations such as data from meteorology or terrestrial magnetism or solar activity[31] and gave an enlarged version of it to the German Physical Society on October 23 in Berlin.[32] He noted that because the method requires the evaluation of a large amount of integrals of the form

$$I(\vartheta) = \int_0^\infty \overline{F(t)F(t+\Delta)} \cos \pi \frac{\Delta}{\vartheta} d\Delta,$$

he proposes to construct a mechanical integrator that would even calculate integrals

$$\Theta(\Delta) = \int_0^T F(t)\, \Phi(t+\Delta)\, dt.$$

Here F and Φ are empirically given functions. If there is no causal relationship between them, Θ is independent of Δ; if there is a causal relationship, Θ depends on Δ and has an extremum, which would help specify this relationship.

Adolf Schmidt, director of the Meteorological Observatory at Potsdam and an inventor, was present at the Berlin lecture and explained to Einstein that the basic idea of the lecture had already been developed by him earlier.[33] Einstein succeeded in getting hold of Schmidt's paper on a planimeter,[34] which was capable of integrating functions of the form $\int y\,dx$, and envisaged developing it into a device for integrating $\int y_1 y_2\,dx$, that is, the integral that Einstein needed.

Einstein grasped the opportunity and, going into technicalities, proposed another idea. Suppose we have a device where the distance of the friction roll from the center of its counterdisk is proportional to y^2. With such an arrangement, integrals like $\int (y_1+y_2)^2 dx$ and $\int (y_1-y_2)^2$ as well as their difference $4\int y_1 y_2\,dx$ can be calculated. Even though he confessed that he was "a low-grade dilettante," he proposed to Schmidt to continue their discussion. Schmidt invited him to Potsdam.[35] Whether Einstein visited him is unknown. He did not return to planimeters.

The Cat's Back Airfoil

"What accounts for the carrying capacity of the wings of our airplanes and of the birds soaring through the air in their flight?" Einstein asked in his lecture on an elementary theory of water waves and flight, delivered to the German Physical Society on June 2, 1916.[36] "There is a widespread lack of clarity on this question. I must confess that I could not find even the simplest answer anywhere in the specialized literature." Then he considered the case of a solid wall or surface with a bulge on it in a flow of a fluid and gave a simple explanation based on Bernoulli's law: the lower side of the bulge is a local increase of the cross section for the flow under the surface. Consequently the flow slows down, and the pressure of the air on the bulge increases; for the flow over the surface the opposite applies. These add up to a net force lifting the surface, for example, the wing of an aircraft (fig. 4.2).

Einstein's statement, that even the simplest answer could not be found in the specialized literature, is surprising. The journal *Zeitschrift für Flugtechnik und Motorluftschiffahrt* (Journal for Flight Technology and Dirigible Flight) had been published in Germany since 1910 under the scientific editorship of Ludwig Prandtl, director of the Göttingen Model Test Station for Aerodynamics (MVA) and professor at the University of Göttingen. The famous Russian expert Nikolay Evgenevich Zhukovsky was also among the editors. In its first volume, Zhukovsky published a paper on his method of how to calculate

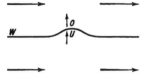

Figure 4.2. Bernoulli's principle.
Albert Einstein, "Elementare Theorie der Wasserwellen und des Fluges," *Die Naturwissenschaften* 4 (1916): 509–10. Courtesy Albert Einstein Archives, The Hebrew University of Jerusalem.

airfoils (the so-called Zhukovsky profiles);[37] in the following years Prandtl and his colleagues explained lift and drag in various types of flight and aircraft. Prandtl even published a chapter on his investigations in a handbook.[38] In the official publication of the Bavarian Academy of Sciences, Wilhelm M. Kutta had presented his calculations on the lift of fluid flow in 1910 and 1911.[39] These papers discussed mathematical methods leading to results that could be compared with experimental findings in wind tunnel measurements and actual flights.

In Einstein's defense, I must add that technical publications on flight came to a sudden stop in 1914 with the outbreak of the First World War, that is, two years before Einstein's interest in flight flamed up. The aerodynamic research done during the war in the aircraft factories, at universities, and at Prandtl's MVA was published only after 1917 in the *Technische Berichte*, a series of publications that were classified. Prandtl himself published his groundbreaking two-part paper on an overview of the history and contemporary state of the theory of flight in 1918.[40]

Einstein's paper would have remained one little popularizing article among many others had he not considered his musings worth implementing. Apparently he continued his explorations into novel wing profiles because, on a spring day in 1917, the head of the Testing Department (and also test pilot) of the Air Traffic Company (Luftverkehrsgesellschaft, LVG) in Berlin-Johannisthal, Paul G. Ehrhardt, found a handwritten document on his desk. It looked like an "impressive piece of work," so without reading it through, he sent it over to Arno Schleusner, a consultant in higher mathematics. Ehrhardt could not identify the author, because the accompanying letter had been sent to commercial director Otto Marx. It was a surprise for him when two days later he found Schleusner deep in discussion with a person with "a mane of iron gray hair." It was Einstein.[41]

Why did Einstein approach the LVG?

One of Einstein's earliest biographers, Carl Seelig, offers the answer: "To be competitive [with the aircraft of the Allied powers], the Luftverkehrsgesellschaft Berlin-Johannisthal turned to various scientists to awaken their interest in technological improvements. One of the few who agreed was Einstein."[42]

I don't think, however, that it held appeal for a pacifist, Nobel-candidate Einstein, member of the Prussian Academy of Sciences.

Peter M. Grosz offers another answer. He maintains that Arthur Müller, owner of LVG, approached Einstein for reasons of publicity and hired him formally as a consultant. Because Einstein felt it his duty to do something for the money, he designed an airfoil.[43]

Albrecht Fölsing's guess is that Ludwig Hopf, Einstein's earlier collaborator at the University of Zurich, now working on aerodynamical problems at the Flugzeugmeisterei der Fliegertruppen, an office of the German Air Force in Berlin-Adlershof, awoke Einstein's curiosity about flight.[44]

We have no reliable arguments either for or against these statements. Einstein's letter to Otto Marx that accompanied his "impressive piece of work" could answer his motive, but up till now it has not turned up.

What did Einstein submit to Otto Marx? The sheets with the calculations were Ehrhardt's cherished treasures in his private collection of documents on the early history of flight, until they were destroyed in an air raid on Frankfort in the Second World War.[45] According to his reminiscences, the calculations referred to the curvature of an airfoil that was to have minimum drag and maximum lift at zero angle of attack, under the condition that the frictional resistance of the air and the thickness of the airfoil were ignored. I will explain these concepts later. The calculations used higher mathematics and took several pages.

We can add to this meager information from Einstein's letter to his closest friend, Michele Besso.[46] In it, he answered Besso's question of how to calculate the bulge of the airfoil: "The permissible bulge is given by the stability conditions of the fluid current; because, with a higher bulge, the airflow does not follow the surface, so vortices appear," and he appended a drawing (fig. 4.3).

The letter to Besso is dated May 14, 1916, almost a year before the airfoil was offered to the LVG. Apparently flight had caught Einstein's imagination during the three weeks he spent in Zurich in April 1916, discussing this problem with Besso.

The rib of the airfoil proposed to the LVG is shown in figure 4.4. The photo was taken when it was tested at the MVA in Göttingen.

Figure 4.3. Bernoulli's principle.
Einstein to Michele Besso, May 14, 1916. Courtesy Albert Einstein Archives, The Hebrew University of Jerusalem.

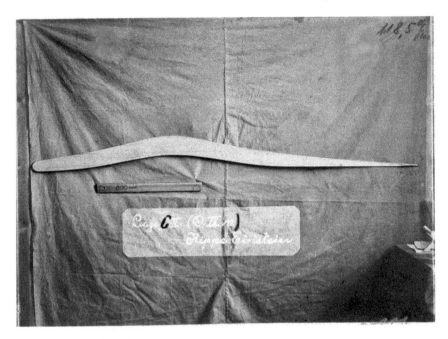

Figure 4.4. Einstein's airfoil rib.
Courtesy Deutsches Zentrum für Luft- und Raumfahrt e. V., Göttingen, Zentrales Archiv, PS 17-5.

Ehrhardt was so impressed by Einstein's calculations that he immediately set about building the airfoil. In the shop, it was nicknamed the "cat's back airfoil."

Ehrhardt's account does not mention whether wind tunnel tests preceded or followed the test flights. The routine procedure in the practical introduction of a new model was that the aircraft company or the inventor submitted it to the Flugzeugmeisterei, which forwarded it to testing in the wind tunnel of the MVA, but this was not always followed. The models were prepared either by the applying firms or by the MVA.[47] The characteristic air velocity used was 9 meters per second (20 miles per hour) in the old wind tunnel.[48]

March 1917 was the time the new institute in Göttingen was inaugurated: a new building with a new wind tunnel, the most advanced in the world at the time.[49] A year earlier a clearinghouse was founded for the results, the Scientific Information Bureau for Aviation (Wissenschaftliche Auskunftei für Flugwesen). Its first publication, the first volume of the *Technische Berichte* also came out in March 1917. The issues were classified, numbered, and confidentially distributed among aircraft companies to keep pace with each other's results, but the names of manufacturers were substituted by numbers. Einstein's airfoil was given the number 95.[50]

Figure 4.5 shows the cross section of a customary wing flying to the left (or, similarly, a stationary wing in an airflow from the left in a wind tunnel). α is called the angle of attack, subtended by the direction of flight v and the chord line C. L is the lift, the upward component of forces acting on the wing, and D is the drag, the component of forces that tend to hinder its forward motion. An airfoil is a good prospect if it has as high a lift as possible with as low a drag as possible. As we have seen, Einstein claimed that his airfoil had these good characteristics even at a zero angle of attack.

What did the model tests prove? We have two polar diagrams that summarize the results. The first was published in the inaugural issue of the *Technische Berichte*, which comprised earlier internal reports from March 15 to October 15, 1917 (fig. 4.6).[51] C_a stands for the lift coefficient: the lifting force L divided by the product of the area of the model and the "dynamic pressure," the pressure calculated from Bernoulli's principle at the given air velocity. C_w is the drag coefficient, also a ratio with the same denominator, but here the drag force D is the numerator. The diagram shows how the two coefficients change with the angle of attack between $-9°$ and $+18°$.

Another polar diagram was preserved by Peter M. Grosz (fig. 4.7).[52] Here the lift coefficient is designated by ζ_a, and the drag coefficient by ζ_w. It is from a technical manual dated March 29, 1917, a work diagram, for it reveals the companies' names: "Fok[ker]," "Einstein (Lvg)," "Alb[atros]," "Eu[ler]," in

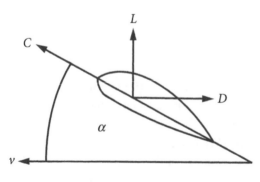

Figure 4.5. Forces acting on a wing.

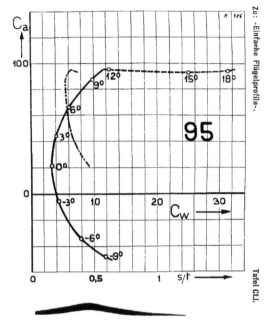

Figure 4.6. Polar diagram.
Max Munk and Carl Pohlhausen,
"Messungen an einfachen Flügelprofilen," *Technische Berichte* 1 (1917):
table CXXXXIX.

contrast to the *Technische Berichte* where, as I mentioned, the airfoils were presented only by serial numbers. Interestingly, Einstein's name is added to the company's name. It must have been treated with special attention.

The pairs of curves indicate the data measured on the lower (*S*) and upper (*L*) surface of the airfoil.

Apparently for reasons of convenience, the abscissa of Einstein's data is put higher than that of the other airfoils. It is striking how soon Einstein's airfoil reaches the angle of attack when lift turns constant or even reverses.

This point of maximum lift is called "stall." An aircraft with Einstein's wing cannot reach a lift coefficient higher than 92, whereas, for example, the airfoil number 94, a Fokker, climbs up to almost 140.

The wing with the Einstein airfoil was mounted on the trunk of an LVG biplane during the next few weeks. Ehrhardt supervised the construction with growing skepticism: he expected that the plane would compensate for the missing angle of attack with a "sagging backwards." "Unfortunately the skeptic in me proved to be right," Ehrhardt continued in his letter to Einstein, "for I hung in the air like a 'pregnant duck' after take-off and could only rejoice

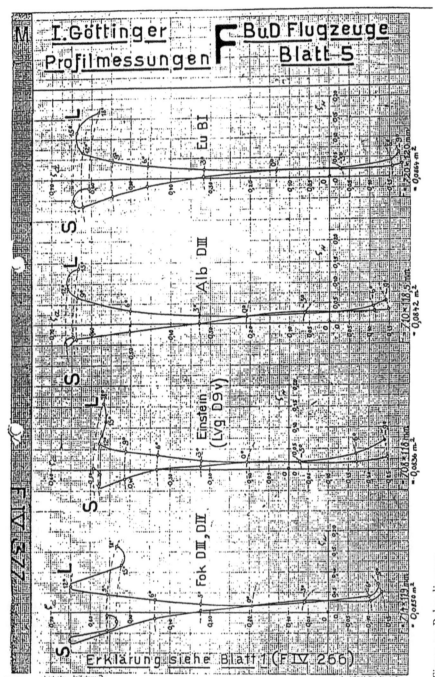

Figure 4.7. Polar diagram.

Peter Grosz, "Herr Dr. Prof. Albert Who? Einstein the Aerodynamicist That's Who! Or Albert Einstein and His Role in German Aviation in World War I," *W. W. I Aero,* no. 118 (February 1988): 42–46. Courtesy Technikmuseum Berlin, Peter Grosz Collection.

when, after painfully flying straight down the airfield, I felt solid ground under my wheels again just short of the airfield fence. . . . The second pilot had no greater success: not until the cat's back airfoil was modified to give it an angle of attack could we venture to fly a turn, but even now the pregnant duck had merely became a lame duck."[53]

Einstein answered Ehrhardt's letter of August 26 in less than two weeks.[54] He started with a straightforward self-criticism: "That is what can happen to a man who thinks a lot but reads little. . . . I have to admit that I have been ashamed of my folly of those days." He repeated his considerations based exclusively on Bernoulli's principle, but he also admitted that, "although it is probably true that the principle of flight can be most simply explained in this way, it by no means follows that it is wise to construct a wing in such a way!" He recognized that "Nature knew well enough why she made birds' wings rounded in front and sharp-edged behind!"

At the end of 1917, Einstein expressed his thanks for an album to Otto Marx: "I was delighted to receive your splendid and clever present. Whenever I look through the beautiful album, I remember with amusement . . . my escapade in the realm of the practical."[55]

The story I have told here started and ended in 1917. We have, however, the oral testimony of another test pilot, Otto Reichert, whom Peter Grosz met with in Germany.[56] He maintained he had flown with Einstein's airfoil—he even claimed that they got unsatisfactory results, and an engine failure put an end to further tests, when the aircraft hit a power line and was destroyed by fire. Because Reichert left the LVG on December 15, 1915, the tests must have been made earlier. How do we reconcile this date with the 1917 of the previous story?

Ehrhardt mentioned a second test with another test pilot "who had no greater success," and another test (or more?) with a wing with a nonzero angle of attack. Was one of the pilots Reichert? Did he leave LVG not in 1915 but in 1917? The latter year looks far more promising a time than 1915. In addition, the interview with Reichert was made in the seventies or eighties of the past century, more than fifty years after the events.

We have a third candidate, too, a certain Hanuschke, flying the biplane. (Maybe he is identical with Bruno Hanuschke, who at the time run a small aircraft-constructing firm in Berlin-Johannisthal.) Carl Seeliger interviewed him, and Hanuschke also mentioned the first pilot, named Eberhard, flying like a "pregnant duck."[57] "Eberhard" is close enough to "Ehrhardt" to assume that he remembered the same first pilot. Hanuschke did not mention a date.

A late echo of this "escapade" comes from a short news note in an Austrian flight journal in 1920: "Einstein and aeronautical engineering. It is announced that the much-discussed scientist Prof. Einstein was once also engaged in aeronautical engineering: in 1916 he had constructed a new airfoil, but the construction could not reach practical utility."[58]

Compasses for Land, Sea, and Air

Imagine a humming top. But unlike your toy top, which will stop after a while, this top is driven by an electric motor. If specially mounted, the top will work as a compass, indicating geographic, not magnetic, north, because it is not influenced by the magnetism of the Earth.

Let us have a closer look at Einstein's share in the development of Hermann Anschütz-Kaempfe's spherical gyrocompass.

It is not easy to follow their alliance. Some gaps in their correspondence are due to loss of letters but also to the fact that Einstein spent weeks in Anschütz's factory in Kiel and their exchange was oral. Nor were minor steps in the development of the gyrocompass documented. There are also limitations to using the Anschütz-Raytheon archives.

Einstein's acquaintance with Anschütz-Kaempfe began in 1915 with his expert opinions discussed in chapter 3. By 1918 their official business had turned into a collaboration and a friendship. In serving as an expert for years on end, Einstein "trained" himself both experimentally and theoretically, even by reading Usener's book on the gyrocompass.[59]

Anschütz 's (and Einstein's) problem was how to prevent disturbances to a gyroscope from the jolts of a ship. The idea was to put the gyroscope in a metal sphere and levitate it by electromagnetic forces. By this method, the magnetic field of magnet wreaths or rings under or around (later within) the sphere will induce eddy currents in its wall when the sphere moves with respect to the magnet. The magnetic field of the eddy currents will, in turn, try to restore the original position of the sphere. In October 1920 Anschütz informed Einstein of two promising experiments.[60] In the first, he put an aluminum hemisphere over three electromagnets; in the second, a whole sphere in between two magnet wreaths, each made up of ten electromagnets. In Figure 4.8 the wreaths are shown in vertical cross section. There is no outer shell there.

This arrangement did not give satisfactory results because of the asymmetric distribution of iron (three gyroscopes) in the sphere. The next arrangement,

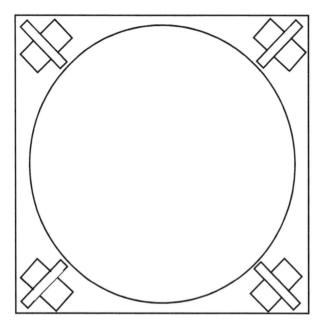

Figure 4.8. Magnet wreath. Vertical cross section.
After Hermann Anschütz-Kaempfe to Einstein, October 10, 1920.

from December, built upon Einstein's suggestion, was an iron ring with wind-ings around it, which was placed under the sphere (fig. 4.9).[61] It gave no better results but was considered promising.

Now we encounter a three-month gap in the correspondence. On March 10, 1921, Anschütz sent Einstein a report, prepared by Karl Glitscher and Max Schuler in Kiel, on a longer effort to cope with problems "with the production of the new ring."[62] This ring magnet was put together from radially arranged metal plates, temporarily separated from each other by oil paper and shellac and held together by a wooden frame.

As we have seen, in December a simple ring had been attributed to Ein-stein. Is this new one with the U-shaped core, a new version of the earlier one, also suggested by Einstein in missing letters, or a new idea of Glitscher and Schuler's? Anschütz's letter sounds like a report to the father of the idea but without saying so explicitly.

The problem encountered with this new ring, again an instability of the position of the sphere, seemed to be insurmountable for Anschütz. He con-sidered, reluctantly, returning to the wreath arrangement of figure 4.8, but

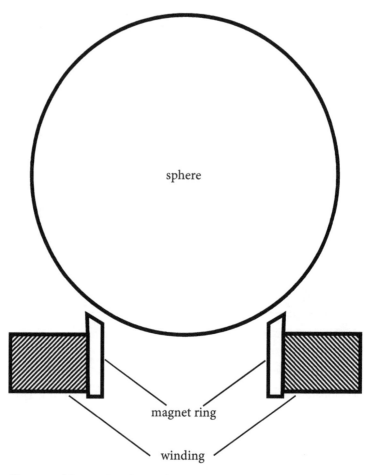

Figure 4.9. Magnet ring. Vertical cross section.
After Hermann Anschütz-Kaempfe to Einstein, December 28, 1920.

this was now made of so many magnets that the asymmetry of the content of the sphere would not entail a problem.

The small electromagnets have alternating polarities to make the summary magnetic field within the sphere close to zero (fig. 4.10). Einstein's first (extant) letter mentioning the ring electromagnet is a reply to Anschütz's letter and Glitscher and Schuler's report.[63] He proposed to double the ring magnet by changing its cross section into a double-U form (fig. 4.11), with windings S_1 and S_2 in opposite directions in the grooves to have alternating south (S) and north (N) polarities.

There is again silence in their correspondence until July, when Anschütz reported that the sphere containing the gyroscopes was put in an outer sphere and floated in a special liquid. It is highly probable that the arrangement had been made earlier, but no mention is made of them in their correspondence. Anschütz added that the two coils proposed by Einstein had been laid.[64] It is all the more surprising then that at the end of the month Einstein returned to the U form by proposing an alternating arrangement of polarities without much explanation.[65]

In August, Einstein visited in Kiel. Ten pages of calculations and sketches that Anschütz's wife, Reta, preserved are the only testimonies to what they

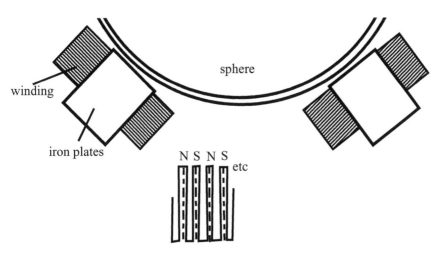

Figure 4.10. Magnet wreath. Vertical cross section.
After Hermann Anschütz-Kaempfe to Einstein, March 10, 1921.

Figure 4.11. Double U-shaped ring magnet.
Einstein to Hermann Anschütz-Kaempfe, March 13, 1921. Courtesy Albert Einstein Archives, The Hebrew University of Jerusalem.

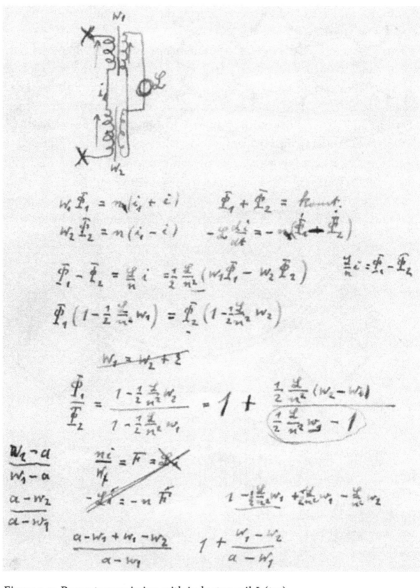

Figure 4.12. Power transmission with inductor coil *L* (*top*).

Einstein's sketches and calculations, Kiel, August 1921. Schleswig-Holsteinische Landesbibliothek, Kiel, Zg.-Nr.: 57/1992. Courtesy Albert Einstein Archives, The Hebrew University of Jerusalem.

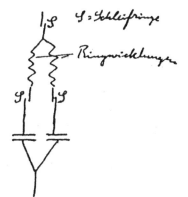

Figure 4.13. Power transmission with sliding rings. S = Schleifringe = sliding ring; Ringwicklungen = ring windings.

Einstein to Hermann Anschütz-Kaempfe, September 18, 1921. Courtesy Albert Einstein Archives, The Hebrew University of Jerusalem.

discussed.[66] In one of them, Einstein sketched two double-U-shaped ring electromagnets around the sphere, but in a month he returned to the idea of simple U form,[67] which Anschütz accepted.[68]

Einstein also worked on other parts of the compass.

He kept pondering on how to feed the magnet when mounted in the inner sphere. To supply the gyroscopes with electricity inside the inner sphere through the water in which it floated was another challenge. He proposed transmission by inductance using oscillating circuits, first with an inductor coil (fig. 4.12), and then, after discovering an error in sign in his calculations, with a capacitor.[69] Another suggestion was to use sliding rings (fig. 4.13).[70]

Einstein even delved into technicalities, including how to put together a ring from small plates by framing them in a china ring (fig. 4.14)[71] and how to make the inner sphere electrically conductive by coating it with a metal more noble than aluminum at the location of the electrode and saturating the liquid in which the sphere floats with a salt of this metal.[72]

Apparently Einstein did not play an active role in the development of the ring magnet in the following years, but Anschütz informed him from time to time about the progress. He announced that he had put the ring magnet within an inner sphere;[73] that he had succeeded in balancing the effect of the higher strength of the geomagnetic field when the ship cruises close to the equator by regulating the distance between the magnet and the inner sphere;[74] and that he had substituted the ring magnet with a very flat coil embedded in the lower pole of the inner sphere.[75]

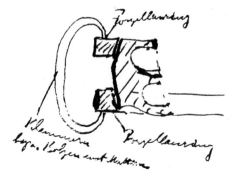

Figure 4.14. Temporary framing. Porzellanring = china ring; Klammern bezw. Bolzen mit Muttern = clamps and bolts with nuts.

Einstein to Hermann Anschütz-Kaempfe, March 13, 1921. Courtesy Albert Einstein Archives, The Hebrew University of Jerusalem.

Problems were met also with the follow-up mechanism of the gyrocompass. As I had already mentioned, the gyroscopes were placed in a sphere that floated in another metal sphere. But how could the north direction of the inner sphere be made visible for the outside observer? Anschütz invented the following mechanism.

Figure 4.15 shows the cross section of the compass spheres from above. Two contacts in both the inner and the outer spheres face each other when both the inner and the outer sphere looks toward north (*left*). When the ship makes a turn, the contacts will not face each other, and the intensity of the current flowing through them and the liquid in which the inner sphere floats will drop because of the higher resistance (*right*). This change is then converted into a signal that controls a reversible motor that moves the outer sphere until a maximum in the current intensity is reached, that is, until the outer sphere follows the inner one in showing toward north.

In 1925 Einstein proposed another arrangement that could also be applied outside the narrow field of gyrocompasses (fig. 4.16).[76] Three stators with three-phase coils are connected in series. They are also magnetically connected so that identical magnetic fields are induced in all of them in the following way.

Let us consider the first. With a special arrangement, a uniform rotary magnetic field will develop. If we put an iron armature in it pointing in a certain direction, the magnetic resistance is reduced, and the flux of the field will be stronger in the direction of the armature than perpendicular to it. This flux will be transmitted to the second and third stators through magnetic coupling. If they also have armatures, they will set themselves parallel to the first—that is, the position of the first armature will be transferred to the next ones.

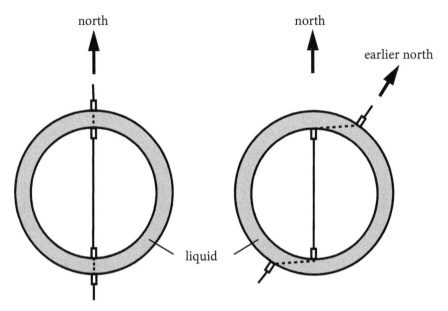

Figure 4.15. Anschütz's follow-up arrangement.

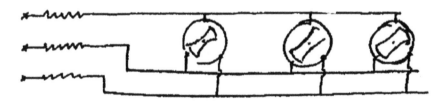

Figure 4.16. Transmission of indicator position.
Einstein to Hermann Anschütz-Kaempfe, August 31, 1925. Courtesy Albert Einstein Archives, The Hebrew University of Jerusalem.

He closed his letter with a concise description of the idea: "A rotary field is influenced by magnetic armatures or also short-circuit armatures in such a way that its strength varies in different directions. Several rotary fields (stators) are coupled electrically or magnetically in such a way that the anisotropy of the magnetic field, generated by the armature, generates a magnetic anisotropy of the other stators, through which a preferred position of the armatures is created."

It sounds like a patent claim. Indeed, this proposal was considered a candidate for an invention, and a contract was signed with "Giro" to perform tests and apply for a patent.[77] In two weeks, Einstein asked Wolfgang Otto, manager of the Kiel factory, how things were going. He was afraid the Kiel people had lost their enthusiasm because he added: "I do think now after all that my suggestion has merit."[78] Otto gave a polite reply. Nothing had been done because of more urgent tasks, but they would make a try "at the first breather."[79] They feared, however, that the arrangement would have low efficiency, "because the voltage is mostly wiped out by the throttles."

Anschütz kept more than one iron in the fire, because the spherical compass was not the only device he was working on. In 1921 he returned to his old idea of making a so-called measuring compass for mining, and Einstein could not help but plunge into the problems connected with it.

This device, later called a gyro-theodolite, was used as a conventional theodolite in setting directions in building roads, tunnels, and mining shifts. It was, however, not influenced by the magnetism of the environment. The sought-for direction is established by giving its deviation from the geographic north (or south) direction with high precision. A mirror is mounted on a gyrocompass, and the beam of a light source reflected from it is observed on a scale through a telescope. In field conditions—where the heavy machinery working in the neighborhood shakes the theodolite and the light spot on the scale with it—it is of utmost importance to make the mirror as independent of the gyrocompass as possible without losing its perpendicularity to the axis of the gyro.

Anschütz mounted the mirror on the axis of the gyro and made it rotate, apparently to avoid the error due to a lack of its complete perpendicularity to the axis. This way, the reflected beam would describe a spot larger than if there was perfect perpendicularity, but with the exact value at its center.[80]

In the spring and fall of 1923, Einstein participated in the development of the device in Kiel. "Apart from a walk with Anschütz, I have not left my hut and usually work far into the night," he wrote to his wife, Elsa.[81] In September, he reported to Anschütz that the reflecting surface of the mirror of the gyro-theodolite is not smooth enough (with an accuracy of 10^{-4} millimeters) to give a nice optical image of the thread; therefore the devices should be sent to Zeiss Optical Works for adequate polishing.[82]

In 1925, on his way from a lecture tour in South America, on board the ship, Einstein met a man from Kiel who reported how the problem with the mirror had been solved by Glitscher by using a floating sphere for the gyro to avoid shocks.

Einstein could not wait for his arrival in Berlin. He sent a letter to Glitscher from Bilbao[83] and proposed another arrangement (fig. 4.17). The gyroscope has a horizontal axis of rotation. The mirror is fixed at its axis. The gyro itself is mounted on a log joint, the axis of which is in rapid forced oscillation to avoid systematic interferences.

Anschütz was amazed by Einstein's curiosity and appetite to solve technical problems. Einstein is interested in the spherical compass, Anschütz wrote to Arnold Sommerfeld, "and works with such enthusiasm on all the tricky questions this unusually audacious construction brings with it, that I can think of nothing better than to be able to come to him at any time with my concerns."[84] A bit of jealousy, however, was not missing. "I'm happy because Einstein has proved wrong again," he wrote to his wife, Reta, in connection with another problem. "I wonder what he will say when he catches sight of this experiment. I think he will again make completely amazed eyes, round as a ball, as he can often make for a plum dumpling or for a [piece by] Mozart."[85]

We have sparse information about how Anschütz's engineer colleagues appreciated Einstein's contributions. When Einstein proposed a special arrangement for the magnet ring, Karl Glitscher remarked to Anschütz: "Einstein the Zionist always enters the promised land, while we only see it from afar."[86] (By the way, Einstein called Glitscher "a very kind and intelligent physicist.")[87] Schuler's opinion must have been different. I have mentioned Einstein's discussion with him on the rotating heated cylinder and on the

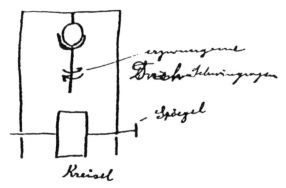

Figure 4.17. Mining gyro. erzwungene Schwing-ungen = forced oscillations; Spiegel = mirror; Kreisel = gyroscope. Einstein to Karl Glitscher, May 27, 1925. Courtesy Albert Einstein Archives, Hebrew University, Jerusalem.

transfer mechanism. I do not dare call it a dispute, for it was a polite and careful exchange (possibly because Schuler was not yet a professor), but it may have been felt by either of them as a quarrel. Schuler was an electrical engineer, the actual head of the Kiel factory, whereas Anschütz had a doctorate in art history, a congenial inventor but only a hobbyist engineer, most of his time spent meditating in his private laboratory in Munich. Schuler played a key role in construction, wrestling with the daily problems. He was the best authority to give an account of the development of the gyrocompass, and he did so in 1962—with no mention of Einstein.[88]

Einstein's name always highlighted anything connected to it. Schuler must have had strong motives for omitting it. Was it personal jealousy? The scourge of being ranked by Anschütz second to Einstein? Or was it a strong opinion that Einstein's involvement was unimportant, or even dilettantish? Again, these are questions that must be left unanswered.

Let us ask another question: Why did Einstein spend time on the gyrocompass? Was it to show and prove his expertise in delicate details of a device? Was it to enjoy the honor and respect of Anschütz? I think so.

He was, however, also attracted by the Anschützes' peaceful, rural, but luxurious life.

His first visit to Kiel was official, only a few days in 1915, when he examined the stability test of the Sperry compass for the legal proceedings between Anschütz and Sperry. It was four years later, in early 1919, when he was invited by Anschütz to his Munich home, apparently to stay there while lecturing at the University of Munich. He was, however, too tired to accept the invitation,[89] so it took another two years before he met Anschütz and his wife at their home in Munich.

Anschütz had a house in Kiel, too. In the fall of 1920, Einstein gave a popular lecture at the Kiel Autumn Week for Arts and Sciences. Anschütz met him at the railroad platform. "We puttered away from the train station in Anschütz's motorboat up to a pier that belongs to the Anschützes' villa," Einstein wrote to Elsa.[90] "It is set right near the water on a small knoll in the middle of a splendid garden. I was then led up to the attic of the villa, where there is an attractive little apartment to lodge visitors; it consists of two small, most tastefully furnished rooms with all the conveniences that the heart could desire and has a splendid view of the Kiel bay. Breakfast is also brought there, so I am surrounded by a matchless tranquility and don't even notice that I am a guest. In addition, Mr. Anschütz and his wife are quiet and content people who haven't the slightest notion about what it means to hurry and scurry about."

Reta Anschütz played the role of his "surrogate-mother,"[91] and he was not against her "motherhood," for Reta was "still very young, pretty, more body than mind."[92] He changed his mind, however, after having spent more time with the Anschützes. "Now I like also his [Anschütz's] wife far better," he wrote to Elsa in April 1923; "she is not at all superficial, and goes excellently with her husband."[93]

He was intoxicated by this style of living. "Think of the house and the sail-boat," he wrote to Elsa in 1920. "We have to create a more human existence for ourselves as well, for all the rural simplicity. There is something fine about a life of meditation. This is most impressively set before my eyes now. Berlin is nerve-racking and deprives me of the possibility of quiet contemplation."[94] The Einstein who wrote these lines was famous for not paying attention to the most elementary necessities of life.

At this time, he and Anschütz began discussing the gyrocompass.

In December 1920 Einstein and his wife received an invitation to spend a longer time with the Anschützes in Munich the next summer. An earlier opportunity arose, however, when Sommerfeld invited Einstein to lecture at the University of Munich. Anschütz offered his home to Einstein and Elsa and, as bait, added that "your room and the music room with the organ are waiting for you."[95] Music was part of the fun in Munich, and apparently the organ was a special attraction.[96] Einstein again had to say no for lack of time.[97] Anschütz repeated his invitation, now to Kiel, for the summer with another bait: "The sailboat or ship is prepared for the summer."[98]

This time Einstein accepted the invitation and in August 1921 spent some time in Kiel, not with Elsa but with his sons, Hans Albert and Eduard.[99] His older son, the nineteen-year-old Hans Albert, made such a deep impression on Anschütz that he "talked in all seriousness about handing over his factory to Albert."[100]

In the second week of July 1922, Einstein visited Kiel with Elsa,[101] not so much to enjoy the hoped-for splendid hospitality as to escape from Berlin. On June 24, 1922, Minister of Foreign Affairs Walther Rathenau was assassinated by rightists in the streets of Berlin. Einstein had been Rathenau's acquaintance since 1915. At the time when Rathenau had accepted this position, Einstein thought it was wiser for a Jew to withdraw from public life, given the strong anti-Semitism of the educated layers of German society.[102] Now, following his own advice and admonitions by "persons to be taken seriously" not to stay in Berlin and not to appear in public anywhere in Germany,[103] he seized the opportunity of Anschütz's repeated invitations. He also withdrew from the Committee on Intellectual Cooperation of the League of Nations and pondered resigning the directorship of the Kaiser Wilhelm Institute of Physics and continuing his life somewhere as a private man.[104]

In Kiel, Einstein confessed to Anschütz that he "is tired of Berlin with everything that goes with it in the way of visits and official things and wants to go . . . into the technical side of things." Delighted as he was, Anschütz was also frightened, "for it is no small matter to stand before all the world as taking a big gun away from more important work."[105] "The prospect of a genuinely normal, natural life in tranquility, connected with the welcome practical employment opportunity in the factory enchants me," he wrote to Anschütz on July 12. "Add to that the wonderful countryside, sailing—enviable."[106] He asked Anschütz "if he could be of value" in the factory.

He was even considering buying an old house there with a wild garden and isolated rooms for himself. It was a villa of historical importance, owned by the heirs of Princess Henriette Esmarch, aunt of Empress Auguste Victoria, wife of Emperor Wilhelm II. But the villa's historical character was an obstacle to buying it, because "the Kiel citizens would perceive the purchase by a Jew of such a historically so heavily encumbered building as a provocative act and take revenge on me somehow."[107]

Einstein also realized that a country life would be too much of a burden for his wife. Elsa sagaciously added that it is easier to hide in a big city than in a small town. The bucolic dream of a peaceful country life gave way to a sober wish to have one or two rooms near the Kiel factory and to spend a considerable part of the year there.[108] Anschütz managed to find a little cottage called Diogenes Ton also in the villa district of nearby Hamburg, full of flower boxes and with a roof slanting low.[109] In the end, however, he thought it more reasonable to offer Einstein a guest residence in the building where he resided when visiting Kiel, with a view to the water and with a sailing boat at Einstein's disposal. The apartment had two entrances: "one for ladies and gentlemen, and one for delivery men and experimental physicists."[110] I wonder which entrance was to be used by Einstein the practical.

So much for the temptation of a luxurious rural life.

A further reason to work for Anschütz could have been money.

Anschütz was a businessman clever enough to keep Einstein's enthusiasm high and to tie this big gun to himself, both as a bogeyman and as a prestigious name for publicity, not only with hospitality but also with financial remuneration. I am not talking about honoraria for expert opinions but about "gratuities" for his contributions. In December 1920 Anschütz proposed to Einstein that they discuss in person Anschütz's obligation to pay him.[111] Anschütz proposed to give, "for the time being," 20,000 marks cash in hand, to avoid taxes. (Einstein's annual salary from the Prussian Academy was 18,000 marks. It was doubled the day after the date of Anschütz's letter.) Further sums followed monthly, for Anschütz reported that "the consignment to Switzerland for the 1st of April is already on its way."[112] In another case, Anschütz assured Einstein that "of course the money will go out punctually to the new address you gave."[113] It is highly probable that the money went to Einstein's first wife and two sons, who lived in Zurich.

In 1923 Anschütz compensated Einstein's services with a "charming bachelor flat that he [Anschütz] furnished here. . . . It is so superb that I regret having to get away from here so soon. It is completely furnished; he bought the furnishings with great love and care as a payment for the previous year; in addition an extraordinary service. . . . It is simply too nice. Besides a view to the woods and water, and the little grand piano! And how nicely everything is made up and how glad he was that he managed to surprise me! I've never thought that external things could make me so happy."[114]

From 1926 to 1938, Einstein regularly received patent fees based on an agreement with Anschütz. There are no data on how many compasses were sold; we know only that a little bit more than $3,000 was transferred to him in seven years between 1928 and 1938. The sums for four years are missing. The annual average of these seven years is $450—not big money, for in 1933, when Einstein was asked how much money he would want to work for the Institute for Advanced Study in Princeton, he proposed an annual sum of $3,000. (Luckily he was given $13,000.)

Their cozy little arrangement came to an end toward the end of 1926. Apparently both he and Anschütz felt the collaboration had come to a point when the legal and financial consequences needed to be settled, and when Einstein paid a visit in Kiel in October 1926, an agreement was drawn up between him and "Giro," Anschütz-Kaempfe's marketing company in the Netherlands, spelling out his remuneration for his share in the development of the gyrocompass.[115]

The agreement defined Einstein's contribution in terms of the claims of Anschütz's patent DE394667.[116] It must have been not an easy task; it was like rendering a colorful but sketchy painting as a sharp black and white drawing. It was all the more difficult because the compass described in the patent was very different from what Einstein had actually worked on.

They agreed that in addition to mentioning Einstein's general collaboration, it was claim 4 that characterized best what Einstein's main contribution was—namely, the U-shaped cross section of the ring electromagnet.

It is, however, also important to note what they did not acknowledge as Einstein's contribution. Claim 3 of the patent, which characterizes the gyrocompass as having "an arrangement of electromagnets whose inducing current is influenced by the momentary position of the gyro system," is not attributed to him. This seems to contradict the widely accepted view that the electromagnetic damping was Einstein's main contribution. I emphasize, however: the answer to the question of which claim to attribute to him and which not must have been the outcome of efforts to subsume Einstein's multifarious ideas, proposals, and hints under preestablished claims.

All in all, the events I sketched do not contradict the results of negotiations compiled in the agreement.

"The old Einstein and the young Einsteins have a bold wish, which is to go to Kiel right at the beginning of the holidays, Albert moved by praiseworthy energy, I by the reclusiveness of the eccentric," Einstein wrote to Anschütz in 1925.[117] Here we read a fourth reason why he was interested in the gyrocompass: he enjoyed being different, not only with his mane of hair, broad-brimmed black felt hat, and the violin under his arm, but also with his adventures outside science.

A final reason may have been to revive his times in the patent office when the hours of work provided him bread and the hours of rest the pleasure of immersing himself in the fundamental problems of science.

Filtering Viruses

Filtering out viruses would be a huge benefit for medicine. But how to determine whether a filter is efficient? A method to determine the pore size of ceramic filters was developed by Einstein and Hans Mühsam, a physician, who treated both him and his Berlin family, and became a lifelong friend.

During their walks, they discussed Einstein's health problems, their biological causes, their treatment, but also Einstein's struggle with the unified field theory. They corresponded up to the 1950s, even after Mühsam had emigrated to Israel.

The lack of a means to determine the efficiency of micropore filters must have arisen during these conversations, probably because Mühsam was interested in sterilization of substances that do not tolerate heat. The method they elaborated and tested was presented by Mühsam at the meeting of the German Microbiological Association and published in 1923.[118]

The size of the largest particle that can pass through a filter is determined by the narrowest diameter of the largest pores. Previous methods were crude. Either water was passed through the pores and its velocity measured, or colloidal solutions containing particles whose size was known (appproximately), were filtered. Einstein and Mühsam proposed letting a liquid wet the pores of the filter, and measure the pressure needed to force the liquid back out. This pressure equals the capillary force with which the liquid adheres to the walls of the pores. According to Laplace's relationship, by measuring this force p, and knowing the surface tension of the test liquid σ, the pore size is $4\sigma/p$.

The authors even built a simple device for this measurement (fig. 4.18). They measured the air pressure when the air started bubbling through the pores of a vessel-shaped filter immersed in ether. They found 6.7 micrometers (0.00264 inches) for the diameter of the largest pores at their narrowest section.

According to Mühsam,[119] the idea was Einstein's.[120] The method, nowadays called bubble-point measurement, has been widely used since that time, but hardly anyone knows that Einstein had a hand in it.[121]

Refrigerators in a Row

Toward the end of 1919, Einstein embarked on a project to develop a cooling process, or as he called it, an "ice machine," together with his colleague, the Nobel Prize–winning chemist Walther Nernst.[122] We do not know much about it. It must have worked only intermittently, for Nernst struggled with compressors and valves during the preliminary tests.[123] They offered the invention to a company in Esslingen, which, however, refrained from developing it

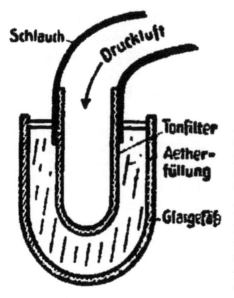

Figure 4.18. Filter control.
Schlauch = hose; Druckluft = compressed air; Tonfilter = earthenware filter; Aetherfüllung = ether; Glasgefäß = glass vessel.

Albert Einstein and Hans Mühsam, "Experimentelle Bestimmung der Kanalweite von Filtern," *Deutsche medizinische Wochenschrift* 4 (1923): 1012–13. Courtesy Albert Einstein Archives, The Hebrew University of Jerusalem.

because of the fire hazard, from which Graff concludes that the refrigerant may have been an inflammable hydrocarbon.[124]

Then they offered it to A. Borsig Locomotive GmbH seeking Borsig's financial and technical help.[125] They declared themselves the owners of a patent on a new cooling process "Nernst/Einstein," proposed a demonstration of its feasibility, and requested 10,000 marks as a compensation for the work already performed as well as further 10,000 marks if the process was used for heat production (so it must have been a reversible process). Borsig was expected to install a test facility in three months. Furthermore, the inventors offered the patent and its use in all types of machines, all over the world, with the exception of the United States and Central and South America. For all these rights, they asked for 250,000 marks. (At that time, Einstein's annual salary was 36,000 marks.) Close cooperation and mutual communication on the further development of the invention was also envisaged. The document, consisting of 1,330 words, is written in heavy "Patentese," with emendations in Einstein's hand. Even though the text mentions a specification of the patent and its explanation attached, they are not available. The only trace that they were doing something practical is Einstein's request from his own institute for 579.85 marks for a mercury beam condensation pump.[126]

On March 4, 1922, Einstein informed his sons, Hans Albert and Eduard, that "the ice machine has made progress. We are just about to complete a contract with the firm Borsig."[127] Its patenting is "completely in the dark," so we are left in doubt whether they had a patent or not. Graff guesses that they applied for a patent, but either the patent office refused it or the inventors withdrew it.[128]

The last news about this venture is in Hans Albert's letter to Einstein. He inquired about the fate of the ice machine.[129] No answer is known, and no patent can be found on Nernst and Einstein's "ice machine."[130]

We have now arrived at Einstein's most famous inventions: the coolers developed and patented together with Leó Szilárd.

"I am sending you the further patent application that I announced," Szilárd wrote to Einstein on September 10, 1926.[131] A letter to the German Patent Office is attached, dated September 13, 1926. The invention was named the "Refrigerator with capillary pump" (no. 2 in table 4.1). The names of inventors were left blank; Szilárd requested Einstein to fill it in "again." Apparently, they had already submitted another application, perhaps no. 1 in table 4.1.

In November 1926 Szilárd offered the Bamag-Meguin company three refrigerator concepts: a water steam jet, a diffusion, and an absorption refrigerator.[132] They have no moving parts, are hermetically sealed, and work continuously without switchover.[133]

For the capillary pump no patent was granted, so we are left with Graff's proposition: capillary forces must have helped pump the absorbent (and the refrigerant absorbed in it) from the absorber to the generator.[134] Graff adds that Einstein may have been inspired by his work with Mühsam on micropore filters, which also made use of capillary forces. One thing is sure: the idea was Einstein's, for he nagged Szilárd even two years later to work on it: "I am happy . . . about the capillary pump," he wrote, "for which I have not yet succeeded in arousing your enthusiasm."[135]

The refrigerators that used a water steam jet had water as refrigerant. They were simple, cheap, and easy to maintain (no. 3 in table 4.1). All the same, Einstein and Szilárd's construction was not patented.

Szilárd and Einstein's next two refrigerators followed a prototype patented by Balthzar von Platen and Carl Munters in 1925 in Sweden.[136]

Table 4.1   Einstein and Szilárd's Common Absorption and Diffusion
Refrigeration Patents and Applications

No.	Patent/ Application	Title	Application date	Granted
1	S73730 I/17a	[Refrigeration invention]	Mar. 1926	No patent
2	S^a	Kältemaschine mit Kapillarpumpe	Sept. 13, 1926	No patent
3	S^a	Dampfstrahlkältemaschine	Oct. 1926	No patent
4	**DE499830**^{b}	**Verfahren zur Kälteerzeugung**	**Oct. 25, 1926**	**May 22, 1930**
5	**DE525833**^{c}	**Verfahren und Vorrichtung zur Kälteerzeugung**	**Dec. 16, 1926**	**May 7, 1931**
	GB282428	Improvements Relating to Refrigerating Apparatus	Dec. 16, 1927	Nov. 15, 1928
	US1781541	Refrigeration	Dec. 16, 1927	Nov. 11, 1930
	CH133906	Verfahren und Vorrichtung zur Erzeugung von Kälte	Mar. 2, 1928	June 30, 1929
6	S^a	?	Dec. 29, 1927	No patent
	FR647838	**Machine réfrigérante avec pompage du liquide par élévation intermittente de la pression de vapeurs**	**Dec. 29, 1927**	**Dec. 1, 1928**
	GB282808	Refrigerating Machines in Which the Pumping of Liquid Is Effected by Intermittently Increasing the Vapor Pressure	Dec. 29, 1927	Not accepted
7	S^a	?	Jan. 24, 1927	No patent
	GB284222	Refrigerating Machine with Organic Solvent	Jan. 23, 1928	Not accepted
8	**DE527080**	**Verfahren zur Erzeugung von Kälte**	**July 14, 1927**	**May 28, 1931**
	GB293865	Improvements in Refrigerating Processes and Apparatus	July 10, 1928	May 30, 1929
	FR671730	Perfectionnements aux procédés et aux dispositifs pour la production du froid	Mar. 19, 1929	Sept. 7, 1929
9	**DE530405**	**Zusatz zum DE527080**	**Oct. 31, 1927**	**July 16, 1931**
10	S^a	Kühlschrankisolierung aus mehreren Lagen Papier	May, 1928	No patent
11	S^a	Kombinierter Kompressor für Kältemaschinen	Sept. 10, 1929	Withdrawn

Note: Patents in bold indicate the original patents; the others are either identical to them or are their combinations.

^{a}Application number unknown.
^{b}Granted for Platen and Munters Ab. Inventors' names from application S76685 I/17a.
^{c}Granted for Platen and Munters Ab. Inventors' names from application S77558 IVb/12a.

In an absorption refrigerator, the refrigerant could be ammonia. It has a low boiling point, −33 degrees centigrade (a little bit below zero degrees Fahrenheit), but under pressure it is kept liquid at room temperature. When a nonsoluble gas (the so-called auxiliary agent) is mixed into the ammonia vapor, for example, hydrogen, the overall pressure in the evaporator stays constant, but because this pressure now consists of the partial pressures of ammonia and hydrogen, each gas behaves as if the entire space were available for it (Dalton's law). The ammonia, under this reduced pressure, evaporates. To put it another way, the ammonia sees the hydrogen as a vacuum in which to expand. Expansion involves taking heat away from the environment. The ammonia and hydrogen mixture enters another vessel, the absorber, where a third agent, for example, water, absorbs the ammonia but not the hydrogen. The hydrogen leaves the absorber and returns into the evaporator, and the water with ammonia, the aqua ammonia, flows into the generator, where it is heated. The ammonia leaves water, is cooled down by passing through an air-cooled condenser, and flows back into the evaporator. The process goes on continuously, while the circulation is maintained by the heat source in the generator.

Ammonia, water, and hydrogen were the original agents proposed by Platen and Munters.

Szilárd and Einstein's refrigerator (no. 4 in table 4.1) had already been patented in October 1926. They resorted to diffusion for separation of the refrigerant and the auxiliary agent. This idea had also been anticipated by another Platen and Munters patent,[137] but they circumvented it by a different arrangement.

The inventors presented five embodiments of the principle. The first is reproduced in figure 4.19. Let us suppose we use ammonia as the refrigerant and hydrogen as the auxiliary agent. The partial pressure of ammonia in evaporator 6 is lowered by bubbling hydrogen from 5 into it, making it evaporate. The ammonia-hydrogen mixture is led into the space between tube 1 and semipermeable vessel 2. The mixture will be richer and richer in the refrigerant, because the hydrogen will leave it through the porous wall of 2, its diffusion velocity being higher than that of the ammonia. The partial pressure of the ammonia will rise until its condensation begins. The condensate returns into evaporator 6 through 9. The hydrogen cools down on 3 and flows back to evaporator 6 through 5. It bubbles through the ammonia, and the cycle continues.

Zu der Patentschrift 499830
Kl. 17a Gr. 8.
Abb. 1
4
3
2
1
10
11
7
8
9
1
5
12
6
13
4
11
5
9
12
3
2

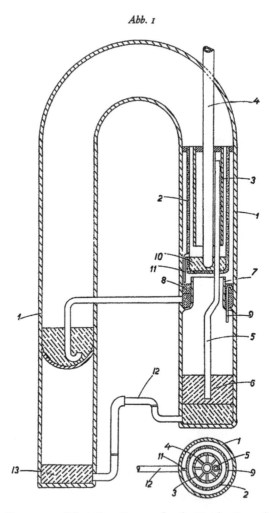

Figure 4.19. DE499830. Process for the Production of Refrigeration. First embodiment

Mercury keeps the process going. It is vaporized in *13* by heating. Its vapor rises in tube *1*, streams along the inner wall of *5*, and condenses on cooling nose *4*. Flowing downward, it drags hydrogen with it, condenses in *10*, passes *11* to *8*, and from there, through *12*, to *13*. In both *8* and in *6*, the condensate collects on top of the heavier mercury.

No. 5 in table 4.1 is another absorption refrigerator. Figure 4.20 shows three embodiments of the German patent. Einstein and Szilárd proposed several agents as the specific novel feature of their invention: (a) butane for refrigerant, water for absorbent, and ammonia, sulfurous acid, or carbonic acid as auxiliary agents; or (b) butane for refrigerant, sulfuric acid as absorbent, and water as auxiliary agent; or (c) methyl bromide for refrigerant, water for absorbent, and ammonia, sulfurous acid, or carbonic acid as auxiliary agents. For other details, the German patent refers to Platen and Munters's "prototype." It is no surprise, for Einstein and Szilárd had sold the application to the Platen and Munters company, and the patent was granted to it.

In the American patent (fig. 4.21), which is the second embodiment in figure 4.20, only butane, water, and ammonia are mentioned as possible agents. Ammonia is bubbled through the liquid butane refrigerant in evaporator 1 through tube 30. The butane evaporates, cools the evaporator, and the ammonia-butane mixture flows into condenser 6 through pipe 5. Water is sprayed in it

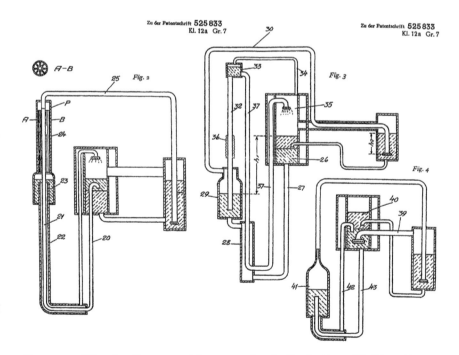

Figure 4.20. DE525833. Process and Apparatus for the Production of Refrigeration

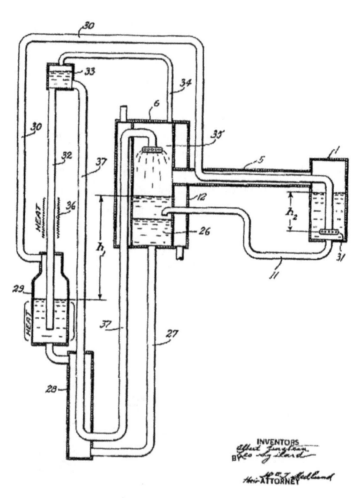

Nov. 11, 1930.   A. EINSTEIN ET AL   1,781,541
REFRIGERATION
Filed Dec. 16, 1927

Figure 4.21. US1781541. Refrigeration

through *37* and *35*, and it absorbs ammonia but not the nonsoluble butane. The partial pressure of butane turns into the entire pressure; consequently, it will liquify and float on the ammonia solution *26*. The butane returns to evaporator *1* through *11*, and the cycle continues.

The ammonia solution flows into generator *29* through *27* and heat exchanger *28*. In it, it is expelled from the water by heating, and leaves through *30* for evaporator *1*. The cycle continues.

The weak ammonia solution is further heated by *36*. The vapor formed in tube *32* lifts the solution into container *33*. On its way, the weak solution flows through tube *37* into condenser *6*. The hot liquid preheats the ammonia solution from the condenser in heat exchanger *26* and cools down. Further cooling takes place in cooling jacket *12*. Then the solution enters condenser *6* to continue the cycle.

In the last decades of the twentieth century, wrestling with air pollution, ozone depletion, and industrial waste, serious interest in absorption coolers-heaters has arisen. An international Heat Pump Programme was launched in 1978 with participation of Austria, Canada, Denmark, France, Germany, Italy, Japan, Mexico, the Netherlands, Norway, Spain, Sweden, Switzerland, United Kingdom, and the United States to work, among many other things, on gas-fired absorption heat pumps. The U.S. Department of Energy started a program in the 1990s, and similar work began in American and European companies. The absorption refrigerator is sailing on these waters under the flag of "Einstein." The sales manager of a British company simply declared that "the origins of today's high efficiency gas-fired absorption heat pump dates back to work carried out on domestic refrigerators by Albert Einstein in the 1920s."[138] In a report on a model of the Einstein-Szilárd refrigerator built according to the American patent by Andy Delano at the Georgia Institute of Technology, the butane-ammonia-water cycle is called the Einstein cycle.[139] Einstein had stipulated in the agreements and contracts that his name not be used for marketing purposes—apparently in vain. There is no contemporary Kechedzhan to invent a telescope with which fainter stars like Szilárd could be observed near the brightness of Einstein's name.

The problem with absorption coolers is that when the pressure in the evaporator is considerably higher than in the absorber, they need a special, expensive pump to move the refrigerant into the evaporator. The difficulty cannot be overcome by substituting the pump with an auxiliary agent. The refrigerator (no. 6 in table 4.1) makes use of a kind of valve that makes a power source for moving the refrigerant back from the absorber to the evaporator unnecessary.

Figure 4.22 shows a section of the pipe, called a receiver, connecting the absorber and the evaporator. In *Fig. 1*, the refrigerant coming from the absorber through pipe *2* has a higher pressure than the pressure prevailing in *1*, because *1* is cooled by water. Valve *4* will open and let the refrigerant in. After a while, heating around *1* is turned on, and the pressure of the coolant vapor will reach a value that is higher than the pressures in *2* and *3*. As a consequence, it closes valve *4* and opens valve *5*. A part of the refrigerant will leave *1* into the evaporator. As soon as the pressure in *1* falls below that of *3* (helped by cooling), valve *5* will close and valve *4* open, and the cycle starts again.

According to another version (*Fig. 2*), two "plugs" of microporous earthenware are substituted for the valves. The pressure differences between *2* and *1*, and *1* and *3*, respectively, will prevent or allow the refrigerant to penetrate the plugs exactly the same way as if the plugs were mechanical valves.

To avoid the intermittent turning on and off of the heater, two further versions are proposed. One of them is shown in *Fig. 3*. The refrigerant is coming from the absorber through pipe *11*, and collects over plug *12*. As soon as the pressure in receiver *13* falls below the pressure prevailing in pipe *11*, their difference will press the refrigerant through the pores of plug *12*, and it will collect at the other plug *14*. As soon as its level reaches *A—B*, the siphon *18* will suck part of it into *16*, which is permanently heated (*17*). After a while, the pressure in *16* will be higher than in *13* and in pipe *15*, which leads to the evaporator; the refrigerant in *11* stops penetrating plug *12*, but the refrigerant in *13* will be pressed through plug *14* until the pressure in *13* falls down because of its being permanently cooled, and the cycle continues.

Ammonia and sulfuric acid, used in refrigerators, lower their partial pressure over the absorbent water substantially, because of their reacting with each other; in addition, absorbents as aggressive as ammonia and sulfuric acid attack parts of the refrigerator. The invention no. 7 in table 4.1 proposes organic homologous liquids, for example, methyl alcohol as absorbent, and octyl alcohol as refrigerant, because they do not react with each other, at least during the short periods they are in contact.

In the absorption refrigerator patented through Citogel Gesellschaft für chemische und technische Erzeugnisse m. b. H., Hamburg (no. 8 in table 4.1, fig. 4.23), alcohol is the refrigerant and tap water the power. Einstein and Szilárd compared their invention to another invention of Clemens Bergl and Walther Dietrich.[140] This was not an absorption process in the strict sense of the word but an evaporator process, as it was later called. The difference between the two inventions is that in Szilárd and Einstein's refrigerator the al-

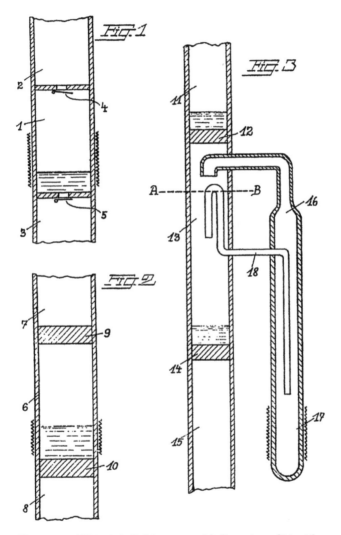

Figure 4.22. FR647838. Refrigerator with Pumping of Liquids
by Intermittent Increase of Vapor Pressure

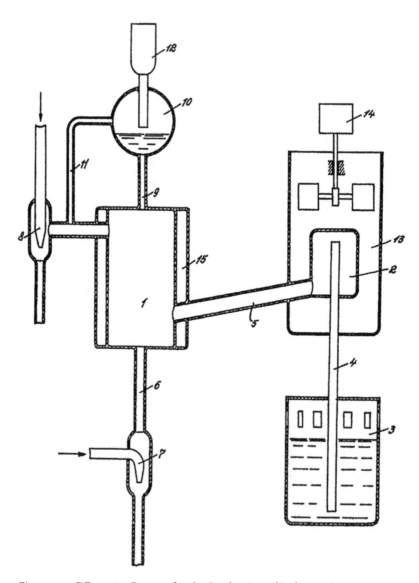

Figure 4.23. DE527080. Process for the Production of Refrigeration

cohol vapor is not sucked by the water jet pump directly from the evaporator but from a large absorber between the evaporator and the jet pump, where the majority of the alcohol is absorbed by water; thus it is not water and alcohol vapor but alcohol dissolved in water that enters the jet pump. The air sucked together with the alcohol from the evaporator is released by another water jet pump.

The methyl alcohol refrigerant is in vessel 3, from which it is pressed into the evaporator 2 by atmospheric pressure. From 2 the alcohol flows through 5 into 1, where it is dissolved in tap water coming from 10 under atmospheric pressure, then leaves through 6 and water jet pump 7 for the sink. Pump 8 sucks the air and alcohol mixture from 1. The vessel containing the evaporator aids ice making, and water turbine 14 helps cooling. The inventors emphasize that although the closed-cycle absorption processes entail sophisticated construction, the present arrangement is simple and needs only common tap water for the jet pump to work. There is no heat source as with conventional absorption coolers. The refrigerant is not recovered but is discharged in the sink together with the water. It consumed 600 grams of methyl alcohol daily in removing heat equivalent to the melting heat of 2.5 kilograms (5.6 pounds) of ice.[141]

This refrigerator, in a robust (concrete) box, was exhibited at the Leipzig Technological Exhibition in 1928 as a "people's fridge" (*Volkskühlschrank*). It was, however, sensitive to the pressure of communal tap water, which changed from building to building and from floor to floor. In addition, methyl alcohol had become expensive. Because of these drawbacks, the refrigerator was never marketed.[142]

Now we arrive at the refrigerators using electrodynamic pumps that have been considered the most ingenious refrigerators developed by Einstein and Szilárd.

The no. 1 in table 4.2, simply called "refrigerator," uses an electrodynamic pump to compress the vapor of the coolant. Figure 4.24 is its first embodiment.

The pump operates as an old, conventional two-cylinder water pump. The cylinders are filled with mercury, an electrically conducting liquid metal. First pump 1 moves the mercury from cylinder 4 to 5, while valve 16 lets the warm vapor of the refrigerant from evaporator 20 in. When the flow of mercury is reversed, the vapor in 4 is pressed through valve 14 into pipes 15 and 19. It condenses in the air-cooled condenser 18 and flows back into evaporator 20. The same happens with the other cylinder with a phase shift. So far nothing new.

Table 4.2   Einstein and Szilárd's Common Refrigerators with Electrodynamic Pump

No.	Patent	Title	Application date	Publication date
1	DE563403	**Kältemaschine**	**Nov. 12, 1927**	**Oct. 20, 1932**
2	DE554959	**Vorrichtung zur Bewegung flüssiger Metall, insbesondere zur Verdichtung von Gasen und Dämpfen in Kältemaschinen**	**Dec. 27, 1927**	**June 30, 1932**
	AT133386[a]	Verdichter für Kältemaschinen	Dec. 22, 1928	
	CH140217[a]	Kältemaschine	Dec. 21, 1928	Jan. 15, 1933
	FR670428[a]	Machine frigorifique	Dec. 26, 1928	May 3, 1930
	GB303065	Electrodynamic Movement of Fluid Metals Particularly for Refrigerating Machines	Dec. 24, 1928	Aug. 19, 1929 May 26, 1930
	HU102079	Hűtőgép	Dec. 5, 1929	
	NL31163	Werkwijze voor het comprimeeren van den damp an het koudmakend middel in een koelmachine, geschickt voor de toepassing van deze werkwijze	Dec. 27, 1928	Mar. 2, 1931 Oct. 17, 1933
3	DE562040	**Elektromagnetische Vorrichtung zur Erzeugung eines oszillierenden Bewegung**	**May 31, 1928**	**June 10, 1932**
4	DE555413	**Pumpe, vorzugsweise für Kältemaschinen**	**Dec. 3, 1928**	**Jul 7, 1932**
	AT133386[a]	Verdichter für Kältemaschinen	Dec. 22, 1928	Jan. 15, 1933
	CH140217[a]	Kältemaschine	Dec. 21, 1928	May 3, 1930
	FR670428[a]	Machine frigorifique	Dec. 26, 1928	Aug. 19, 1929
	GB344881	Pump, especially for Refrigerating Machines	Dec. 8, 1929	Mar. 3, 1931
5	DE565614	**Kompressor**	**Sept. 10, 1929**	**Nov. 17, 1932**
6	DE556535	**Pumpe, vorzugsweise für Kältemaschinen. Zusatz zum DE555413**	**Apr. 14, 1930**	**July 21, 1932**
7	DE561904	**Kältemaschine**	**Apr. 14, 1930**	**Sept. 29, 1932**
8	DE562300	**Kältemaschine**	**Apr. 14, 1930**	**Oct. 6, 1932**

Note: Patents in bold indicate the original patents; the others are either identical to them or are their combinations.
[a]Patents including both DE554959 and DE555413, with minor modifications.

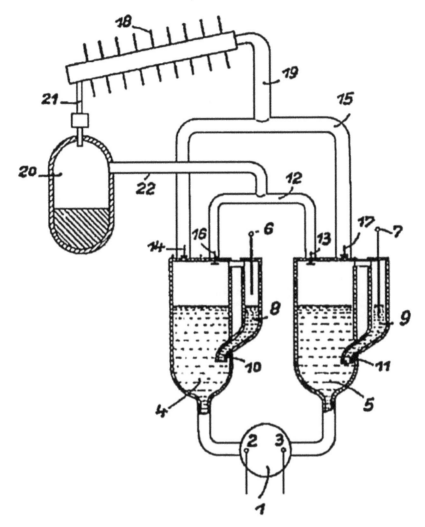

Figure 4.24. DE563403. Refrigerator

The devil is in the pump. There is a narrow rectangular slit *36* in the pipe connecting *4* and *5* at their bottom. It is shown in cross section in Figure 4.25. Sides *34* and *35* are electrodes through which electric current is sent through the mercury, perpendicular to the pipe.

Now an electromagnet is applied in such a way that its magnetic field is perpendicular to both the electric current and the pipe. In this case, the so-called Lorentz force acts on the mercury. The Lorentz force *F* is perpendicular

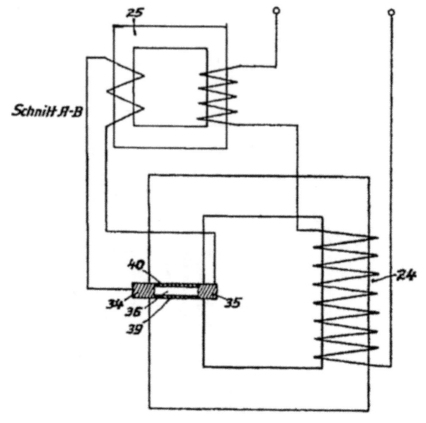

Figure 4.25. DE563403. The electromagnetic pump

to both the electric current *I* and the magnetic field *B* (fig. 4.26). It is parallel to the pipe. Its direction is determined by the right-hand rule. The mercury will start flowing.

To change the direction of the flow, we need only to change polarity. This is done automatically by small electrodes built into side tubes 8 and 9 of figure 4.24. They are parts of auxiliary circuits. When the mercury level sinks below a certain level, the needles cannot touch it, the circuits are alternately interrupted, and they trigger a change of polarity. These tubes communicate with the cylinders through narrow openings 10 and 11 that delay the inflow and outflow of the mercury, and so prevent the "fibrillation" of the polarity change.

When using alternating current, the electrodes can be dropped, for the electric current can be induced in the mercury itself by the changing magnetic

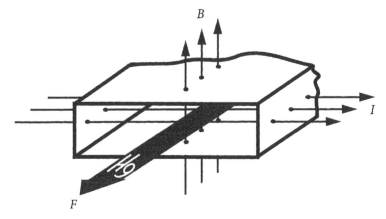

Figure 4.26. The Lorentz force

field—an idea fully made use of in the invention submitted a month later (no. 2 in table 4.2, fig. 4.27).

A liquid metal, for example, mercury, is to flow in a thin annulus *3* between core *2* with radially arranged laminations in it and pipe *1*. Pipe *1* is also surrounded by laminations (e.g., *8* and *9*), arranged radially along it. Coils *4* are wound through the laminations to make them electromagnets.

By shifting the phase of the electric current in subsequent coils *4–7*, a moving magnetic field ("gliding field") is produced that induces current loops in the liquid metal with their planes perpendicular to the magnetic field. The magnetic field and the electric current induce a Lorentz force that moves the mercury downward.

In their next invention, no. 3 in table 4.2 (fig. 4.28), Einstein and Szilárd showed how to construct a motor for producing a to-and-fro motion without switchover.

It consists of two circuits. They work on the principle of the single-phase asynchronous motors—namely, they produce forces that always act in the direction of their actual motion.

Cylinder *9* moves inside standing part *10*. There are shortcut coils around it. The coils in *10* are laid in grooves *a*, *b*, *a'*, *b'*, and so on, *a* and *a'*, as well as *b* and *b'* playing the same role in the motor's operation. The magnetic field of the stator consists of a field with upward direction and another one downward; thus the net force acting on *9* is naught when it is at rest. If *9* is

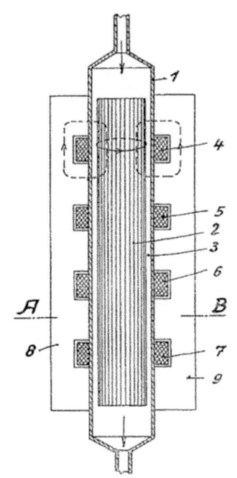

Figure 4.27. DE554959. Apparatus to Move Molten Metal, Especially for Compressing Gases and Vapors in Refrigerators; figure 1

moved upward, a force appears that will lift it until the reaction of spring 23 stops it, then pushes it downward. Now the magnetic force will help the downward motion until the spring below stops it and pushes back. The oscillatory motion continues.

If the device is used as a pump in a refrigerator, it can circulate vapor from the evaporator to an air-cooled condenser and back.

In invention no. 4 in table 4.2, the embodiment coincides exactly with that of no. 2, but the working fluid is not mercury but alkali metals and their alloys. These have lower density, consequently less friction in turbulent motion, and lower loss of energy. (Apparently the inventors encountered this

Zu der Patentschrift 562040
Kl. 21d² Gr. 18

Abb. 2.

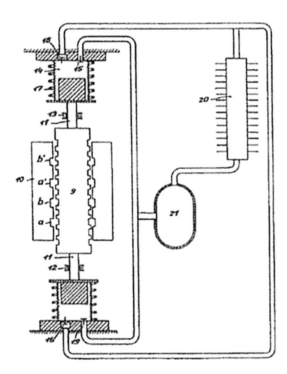

Figure 4.28. DE562040. Electromagnetic Apparatus for the Production of Oscillating Motion

problem during the five months that elapsed after the submission of application for no. 2.)

The applications submitted between December 21 and 27 in several countries are combinations of the two main German patent applications, nos. 2 and 4. They do not mention no. 1 but use it in the text and figures. The only exception is the Hungarian patent, which was submitted in 1929.

With their next compressor, no. 5 of table 4.2, their intention was to make the separation of the working fluid, for example, mercury, from the refrigerant easier. In earlier absorption coolers, the space where the refrigerant leaves the mercury is placed after the gliding-field motor; in this invention, it is placed before it.

The last series of common patents on gliding-field motors was submitted the same day, April 15, 1930.

No. 6 adds a wire net to the earlier constructions to make gas separation perfect.

No. 7 intends to keep heat loss low when the fridge is not in heavy use by keeping the pump with the alkali metal in it just over the metal's melting point when out of use, and keeping it high when it is working, to avoid damage to tube windings because of overheating.

Finally, no. 8 proposes potassium-sodium alloy with 20 percent potassium content as refrigerant, because it is less expensive than pure potassium, and its melting point is lower than that of pure sodium. Apparently they had unsatisfactory results with a refrigerator they built in March that used only potassium.[143]

Along with their common patents, between June 1928 and January 1931, Szilárd submitted (and was granted) nine patents on electrodynamic pump refrigerators. By reading them, we can follow the problems encountered during the development of their common inventions and find ideas that appear in them. The patents propose means of avoiding whirls,[144] unwanted induced electric currents,[145] the backflow of mercury,[146] or its sticking at places from where it cannot flow back into circulation.[147] They propose means of boosting overall efficiency,[148] as well as the efficient separation of cooler gas and the working fluid.[149] They propose improvements on the heat exchanger,[150] the compressor,[151] and the stopcock between condenser and evaporizer[152] and even suggest how to make mechanical assembly easier.[153] The final construction was built of prefabricated elements proposed in this patent.[154]

The development was carried out in the research laboratory of AEG with engineers Albert Kornfeld (later Hungaricized as Korodi) and László Bihaly, Szilárd's fellow students in his years at the Budapest Institute of Technology, under Szilárd's leadership as a consultant. Apparently it was Einstein who made use of his acquaintances to convince AEG to host the venture, for in September 1928 he "roused [AEG] and did it forcefully."[155] Not that he spent much time on it. An acquaintance of his at the AEG actually "roused" AEG while Einstein was reading Spinoza with great joy.

Now that we have a picture of what they achieved in electrodynamic pumping, let us ask the question, What was Einstein's contribution?

There is a remarkable passage in the very first paragraph of their second common patent's description (table 4.2, no. 2): "One can use such a device, e.g., to pour molten metal in mold." It is remarkable because Szilárd had applied for a patent on "a process for pouring molten metals into a mold by using electric currents" in 1926.[156]

The patent makes use of both electrodynamic pumping and the so-called pinch effect.

In its first embodiment, the vertical section (*Abb. 1a* in fig. 4.29) shows the molten metal 1 in a tube; 2 and 3 are electrodes. In the horizontal section (*Abb. 1b*), N and S are poles of a magnet. The Lorentz force moves the metal upward or downward, depending on the direction of the current.

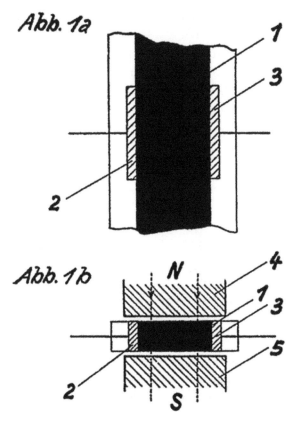

Figure 4.29. DE476812. Process for Pressing Metals in Molds by Means of Electric Currents. First embodiment

In the sixth embodiment, the electric current flows *along* tube 30, within the molten metal (fig. 4.30). Then the magnetic field induced by the current will exert a pressure on the molten metal that falls from the periphery toward the axis of the tube, and this will press the metal from 34 into the mold 33. This is the pinch effect, also a consequence of the Lorentz force.

Szilárd's patent gives five embodiments for the first arrangement and four for the second. It is a reasoned technological description, based on electrodynamic calculations.

Was Szilárd the first to use Lorentz force for practical purposes?

No, he was not. In 1907 a British inventor, Frank Holden, was granted a patent for a "mercury-meter"[157] (intended to meter electricity, not mercury).

A section of tube 7 in which the mercury circulates is a capillary, with dimensions determined in a way to make the flow proportional to the current sent through electrodes 8a and (unseen) 8, both perpendicular to the page (fig. 4.31).

Permanent magnets 2 and 2a are arranged perpendicular to the electrodes. The amount of mercury transported in a certain time indicates the ampere hours and watt hours. The construction can be set to operate continuously or intermittently by including a switch for reversing the polarities.

This is the "classical" electrodynamic motor.

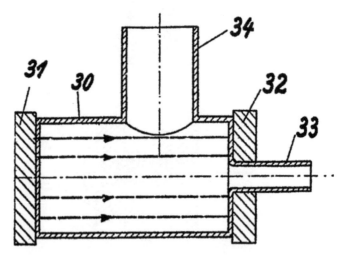

Figure 4.30. DE476812. Process for Pressing Metals in Molds by Means of Electric Currents. Sixth embodiment

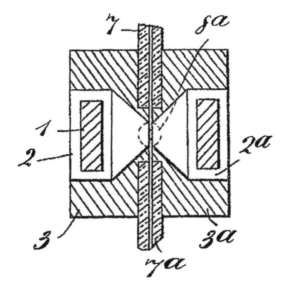

Figure 4.31. Holden's electrodynamic motor (US853789).

There are, however, other predecessors. After a preliminary evaluation of applications for patents nos. 1 and 2 in table 4.2, the engineers of the patent department of AEG where the experiments were being done concluded that a general claim can be made only for a special embodiment but not for the principle.[158] They referred to patents and a technical report that described the idea of moving mercury in an electrodynamic way, as well as the gliding-field motor.[159] AEG even gave preference to another invention on the same principle[160] and submitted it before Szilárd and Einstein's applications.

The idea was in the air. An American, Millard C. Spencer, submitted an application for a "fluid-conductor motor" on July 27, 1927,[161] just four months before Einstein and Szilárd applied for their first electrodynamic patent. Spencer's goal was to pump electrically nonconductive and corrosive fluids, and fluids that interact with lubrication or sealings in traditional pumps. His further goal was to have a pump that worked silently and "which will operate with little attention and in which the fluid connected with its operation may be located in a sealed container without the use of stuffing boxes through which parts need to move into and out of the container, as is desirable in the case of refrigerating machinery which employs fluids the leakage of which would be objectionable."

The pump is an electrodynamic pump already well known to the reader. An electromagnet producing alternating magnetic field induces the current in the mercury. The mercury is pumped from one tank to another through a tube connecting them at bottom. The flow of mercury is reversed when a tank is full by using pins that touch mercury at a certain level. The arrangement is the same as Einstein and Szilárd's first patent: the two tanks make two cylinders in which the mercury works as pistons. The fluid to be pumped is above the mercury.

Another American, Kenneth T. Bainbridge, who became director of the Trinity test at the Manhattan Project, applied for patenting a "liquid conductor pump" on May 28, 1928, six months after Einstein and Szilárd.[162] It is an electrodynamic pump that makes use of the pinch effect.

One thing is sure: the electrodynamic pump was not invented by Einstein, nor was its use in refrigerators, because Spencer also mentioned this possibility.

Nowadays when a complete family of electrodynamic pumps has emerged, the gliding-field motor or pump has a more sophisticated name: "annular linear induction pump" (ALIP).

In the fall of 1930, AEG decided to reduce its research institute by 25 percent. The fate of the gliding-field refrigerator project was to be decided by a commission, and so Szilárd proposed approaching General Electric.[163] He also approached Director Hirst of a research laboratory in England.[164] His next plan was to get Siemens and Brown Boveri interested in the project, with the consent of AEG.[165]

Such talk indicates Szilárd's growing fret. The financial situation in Germany was deteriorating. "I don't know whether we will succeed in completing the construction of our cooler in Europe,"[166] he wrote to Einstein in 1930, and he initiated the patenting of a "refrigeration plant" in the United States under his own name for "clearness" (whatever that might mean), but with the possibility left open to add Einstein's name, because "it will be considered a common property as all the refrigeration inventions."[167] The next summer, however, he was of the opposite opinion and wanted to complete the development in Europe to be able to appear "over there" (in the United States) with the invention ready.[168]

Einstein seized the opportunity of his visiting professorship at the California Institute of Technology in 1931 to send the drawings of an "ejector-type refrigerator" to his old acquaintance, Gano Dunn. Dunn informed him[169] that the idea may have already been patented in the United States by Daniel F. Comstock and attached a newspaper clipping.[170] (It was patented in 1933.)[171]

Dunn was discussing a possible cooperation with Crocker-Wheeler Electric Manufacturing Company. Einstein forwarded the news to Szilárd, who reassured him that Comstock's invention had nothing to do with theirs.[172] It is remarkable that Einstein himself could not see the difference, or else he did not want to spend much time on the matter.

In December 1931, Szilárd made a surprising proposal: let Einstein sign a statement renouncing the coauthorship of ten patent and trademark applications in Szilárd's favor (table 4.3).[173]

Einstein was in Pasadena and answered months later, in July 1932. Referring to his own uncertain future, he "did not consider it expedient" to sign the declaration.[174] Szilárd assured Einstein that the three common fundamental applications (probably nos. 1, 2, and 4 in table 4.2) were not on the list. He apologized for the request by adding that his motive had been to avoid using Einstein's name unnecessarily.[175]

Szilárd followed Einstein's wish not to use his name for commercial purposes, but he did use Einstein's prestige to get AEG interested in the inventions[176] and to help himself get a visa from the Berlin consulate of the United States. Einstein was generous. In his first recommendation, he declared that the purpose of Szilárd's American visit was "to promote our joint work,"[177] and in the next one, he called Szilárd his associate who intended to continue the work in the United States that they had commenced together and in which he himself was interested.[178]

In March 1933 Szilárd gave up German patent 562300 (no. 8 in table 4.2) in his and Einstein's name. The German Patent Office asked Einstein whether Szilárd had been authorized to make this statement on behalf of Einstein.[179] In April the patent office informed Einstein that he was now the sole owner of the patent,[180] indicating that his answer must have been in the negative.

The last mention of a refrigerator patent is in a short exchange of letters between Einstein and a certain Melanie Jaeger in 1934. She was anxious about an article in the *New York Evening World* in which the patent was described in detail (unfortunately I was unable to find it) and offered to mediate between Einstein and an entrepreneur in the United States to patent it in the United States and produce it. Einstein directed her to Leó Szilárd.[181] The interested manufacturer was apparently a certain Julius Janowitz.[182]

Table 4.3   Patent and Trademark Applications of Which Szilárd Requested
Exclusive Authorship

Application number	Title	Date	Patented as
	Patent applications		
S. 85 876 VIII/21 d 2	Oszillierende elektrische Kraftmaschine, im besonderen als Antrieb von Kompressoren bei Kältemaschinen	May 31, 1928	
E. 39 852 I/17 a	Kompressor für Gemisch	Sept. 10, 1929	DE565614
E. 39 853 I/17 a	Kombinierter Kompressor, insbesondere für Kältemaschinen	Sept. 10, 1929	Withdrawn
E. 64.30 I/17 a	Kältemaschine mit Strahlpumpe und Kondensator	Apr. 14, 1930	
E. 40 537 I/17 a	Kältemaschine mit Strahlpumpe und Kondensator	Jan. 22, 1931	
E. 40 538 I/17 a	Pumpe, insbesondere für Kältemaschinen mit besonderer Betriebsflüssigkeit	Jan. 22, 1931	
	Trademark applications		
S. 69 767 /21 d	Oszillierende elektrische Kraftmaschine, im besonderen als Antrieb von Kompressoren bei Kältemaschinen	May 30, 1928	
E. 41 548 /17 a	Kompressor für Gemisch	Sept. 10, 1929	
E. 41 549 /17 a	Kombinierter Kompressor, insbesondere für Kältemaschinen	Sept. 10, 1929	
E. 700 30 17 a	Kältemaschine mit Strahlpumpe und Kondensator	Apr. 14, 1930	

It is difficult to tell whether Einstein had a substantial income from their common patents. In 1927 Einstein and Szilárd agreed on how to share royalties: for all the patents applied for in the past or in the future by both or either one of them in this field, their share will be fifty-fifty. In the case that Szilárd's income sank below the salary of an assistant professor, the net gain would be used to raise his income to that level, Einstein would have the same amount, and they would halve the rest.[183] (I confess I see no difference between the two cases until Szilárd needs more than half of the royalties.) In 1929 Szilárd proposed to share the profit in 1:3:3 ratio, with one-seventh for Kornfeld.[184] Actual sums are barely mentioned in their correspondence.[185]

Magnetostrictive Reproduction of Sound

Einstein's next partner was Rudolf Goldschmidt. As we have seen in chapter 3, Einstein had "worked" for him already in 1921 as a consultant, and he had approached Einstein in 1922 for an expert opinion on a patent case.

They kept in close touch in the following years. They lived not far from each other. From among the ten extant letters from Goldschmidt between 1928 and 1932, seven remember Einstein's wife and five mention invitations for dinner or discussions. If they get together in person often, this would partly explain why their correspondence is so sporadic. After Einstein emigrated to the United States in 1933 and, a year later, Goldschmidt to England, communication was left to their pens. Two air raids on Sheffield where Goldschmidt lived during the Second World War, as well as his moves are responsible for losses of their correspondence,[186] especially letters from Einstein.

A patent that bears both Goldschmidt's and Einstein's names was published on January 10, 1934. It sets out an arrangement for sound reproduction by means of magnetostriction.[187] It had been applied for on April 25, 1929.

What is magnetostriction? If a coil is wound around an iron rod magnet, and an electric current is sent through it, the rod's magnetism changes, but so, too, does its length. The change is rather small because of the rigidity of the rod, which counteracts magnetostriction.

Our inventors' idea was to diminish the rigidity of the rod by keeping it under external pull or pressure, turning its state close to labile (a state where the external pull or pressure is close to its tear or buckling, respectively) and capable of following the sometimes-considerable changes of the electric current in the coil. When this arrangement is used, for example, in a microphone or loudspeaker, the lengthwise vibration of the rod can follow the oscillations of the current in the coil produced by the sound to be transmitted.

In *Abb. 1* of figure 4.32, you can follow the idea. An iron rod *B* is pressed by two screws up to the critical value in buckling. The bending forces keep balance with the rigidity, and even a small excitation by the magnetic field of coils D induced by the electric current flowing in them can elicit considerable changes in the rod's length. Being under pressure, these changes will bend the rod, and the bend is transferred to a diaphragm by an element fixed to it at its midpoint.

Abb. 3 of figure 4.32 shows an example close to the real application. The iron rods B_1 and B_2, together with flexible components C_1 and C_2 make the scales of a balance G, which hangs on a flexible rod M. In contrast to the previous case, here the rods are not compressed but pulled by screw P to achieve their labile state. The rods are magnetized so that when the one dilates under the effect of magnetostriction, the other returns to its original length, thus causing G to tilt. The motion of G is transferred to diaphragm W by O and V. F prevents sideway motion.

In a letter of February 28, 1929, Goldschmidt sent Einstein two patent specifications submitted that day in their names.[188] Einstein expressed his consent to the formulation of the inventions but emphasized that his share should be only 33 percent.[189] There are three typed specifications in the Albert Einstein Archives next to the letter, so the two mentioned in the letter must be among them. One is on magnetostrictive sound reproduction,[190] later patented as DE590783.[191] The second should be one of the two others, both proposing special drives to avoid nonlinearity of magnetic attraction around the magnet's poles.[192]

Curiously enough, on April 9, Det Tekniske Forsøgsaktieselskab (Goldschmidt's past assistant called it Goldschmidt's firm),[193] informed Einstein that two specifications were filed by the German Patent Office under application numbers G. 75753 and G. 75754,[194] neither of them identical to their common patent, G76 240. In addition, simultaneously with the application for the Goldschmidt-Einstein patent on April 25, another application bearing only Goldschmidt's name was also submitted to the German Patent Office.[195] Now we have three dates for the submissions, but even the inventions are difficult to identify. What is sure is that there is only one common patent granted.

What could have been Einstein's contribution? The idea of diminishing rigidity by inducing stresses in a solid body to bring them close to a labile state turned up in all of Goldschmidt's thirteen inventions that he was working on in the second half of the 1920s. In all of them, it is the diaphragm that is subjected to push or pull, never any other part of the microphone. The

Zu der Patentschrift **590 783**
Kl. **21 a² Gr. 1**₀₄

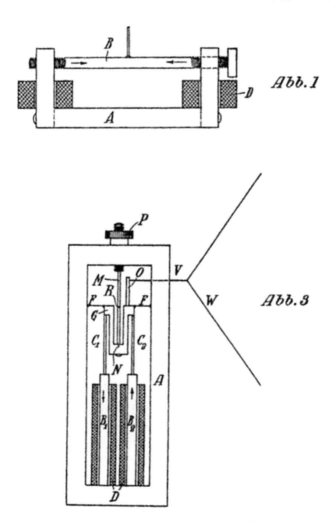

Figure 4.32. DE590783. Magnetostrictive microphones.

Goldschmidt-Einstein patent is the only one that applies the same procedure to a rod magnet to enhance its magnetostrictive vibration. It is tempting to attribute this idea to Einstein, and not to attribute to him the use of stresses to offset rigidity. This second guess is corroborated by Einstein's remark in his next letter to Goldschmidt: "I still fear that the idea of the compensation of elasticity, which is wonderful indeed, is too subtle to translate into practice."[196] It is not his style to praise his own idea as "wonderful," so the compensation of elasticity must be Goldschmidt's idea.

In the same letter Einstein mentioned that "The Tekniske Forsøgaktie-selskab . . . certified to me my proportion of the fatherhood of the egg we are to lay together." (He refers here to a letter from the Tekniske Forsøgsaktieselskab.)[197]

The egg to be laid together was first mentioned in the dedication on Einstein's photo (fig. 4.33) sent to Goldschmidt in November 1928:

> A bit of techno here and there
> Amuses even me to dare
> To ask the brazen question whether
> We two can lay an egg together.[198]

Upon which Goldschmidt replied with the little poem:

> For *one* egg to come from *two*
> Is somewhat difficult to do!
> The best solution I would say,
> That is, if you find it OK,
> We'll each lay eggs upon a bet
> And mix them in an omelet![199]

I must, however, emphasize that there is no definitive correspondence on this patent between Goldschmidt and Einstein, and the egg mentioned by Einstein may refer to any or both of the inventions submitted.

Apparently prompted by Einstein's skeptical remark on the compensation of elasticity, Goldschmidt turned to the research laboratory of the Siemens company. As he reported to Einstein,[200] its head, Gerdin, showed interest in the "labilizer" and seemed to be prepared to carry out tests. Gerdin also added that only specific iron alloys would give adequate results.

The patent office made only formal objections to the patent, Goldschmidt continued his letter (now it is only about a single patent and not about two; apparently the other had already been rejected). He said he would like to discuss it with Einstein, along with a completely new idea in the loudspeaker business, which he would develop together with Einstein.

To lay eggs needs time for both hens and inventors. It is the end product of a process. Goldschmidt turned to Einstein time and again with problems and

Figure 4.33. Portrait with dedication to Rudolf Goldschmidt.
Courtesy Albert Einstein Archives, The Hebrew University of Jerusalem.

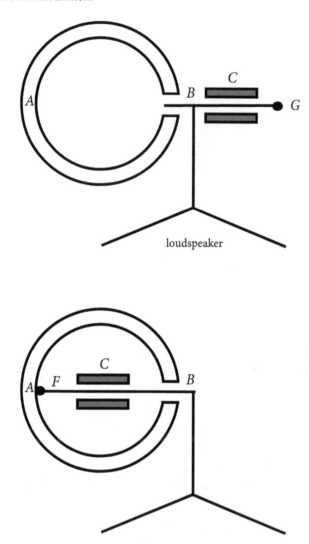

Figure 4.34. Einstein's and Goldschmidt's proposals.
After Rudolf Goldschmidt to Einstein, November 15, 1928.

reported his concerns. Their correspondence indicates a collaboration in which proposals, guesses, and ideas were exchanged on both large and minute details.

To show a few examples, in the first (extant) letter after 1922, on May 2, 1928, Goldschmidt requested Einstein's opinion on a draft of a patent claim that he intended to submit to the British Patent Office.[201] This was on "a diaphragm under magnetic or electrostatic attraction in which stresses are produced artificially . . . , so that the diaphragm is put in a pseudo-astatic equilibrium." Goldschmidt submitted an invention to the British Patent Office on May 29,[202] which described a diaphragm that comes close to that mentioned in his letter. None of the patent claims echoes those of the letter. I dare not say that the reformulation is due to Einstein, but I also cannot say that it is not.

An idea of Einstein turns up in a Goldschmidt letter that is also on sound reproduction but without magnetostriction. Goldschmidt made sketches of both Einstein's proposal and his own version (upper and lower figures, respectively, in fig. 4.34).[203] *A* is a permanent magnet, *B* a tongue, *C* an electromagnet, and *G* and *F* are joints keeping *B* fixed. The inverse Y is a rod ending in a cone. No explanation is given of how the arrangement works. Apparently *B*'s magnetism is changed by the changing electric current in coil *C*. *B* will move up and down in the gap between the poles of *A*, and it reproduces the vibrations of the sound that controls the current. The difference between the two versions is not of principle but of practical realization.

"It would be best to discuss details in person," Goldschmidt proposed, and these were the last words written about this arrangement.

Hearing Aid

"I have not yet been able to formulate what I can say of your very interesting suggestion," Goldschmidt wrote to Einstein in October 1931. "The cranial bone has a relatively high rigidity and high damping. The motoric part of the magnet may also be more massive as usual, as you point out."[204]

With no extant letters, we can only guess that Goldschmidt was writing about the hearing aid that we find in books and papers that mention Einstein's inventions. Their communication continues only after more than a year with sketches of how a handle could be fixed on the cranial bone and advising that a dentist would be best consulted on how to do it.[205] Goldschmidt offered to turn to a professor of dentistry, an acquaintance of his, but kept open the possibility that Einstein or Mrs. Eisner might also have appropriate connections.

Olga Eisner, a singer, and her husband, the concert pianist Bruno Eisner, met the Einsteins in 1928. It did not take long to learn that Mrs. Eisner struggled with loss of hearing, a serious drawback for a musician. Einstein and Goldschmidt decided on the goal of constructing a convenient hearing aid.[206]

Goldschmidt worked enthusiastically. Four days after his first letter, he proposed a layer of dental cement to fasten the disk to the cranial bone.[207] Meanwhile Albert and Elsa Einstein had found a dentist, Professor Axhausen. Now Goldschmidt changed his mind and proposed a simple mechanical sound transfer to the cranial bone.[208] "I would like very much to discuss with you the pros and cons, which it is impossible to do in writing. . . . would you and your wife give us the pleasure of coming to us for dinner or for a cup of tea?"

Upon Einstein's recommendation (which is not extant), the Mendelssohns (apparently the owners of the bank Mendelssohn & Co. that administered the finances of Einstein's institute) offered Einstein 500 to 1,000 marks for help, with the comment that if the venture succeeds, "then you will do mankind an unheard-of good turn" (unheard of not only for those with hearing problems).[209]

Meanwhile, Goldschmidt turned to Lautenschläger (it is highly probable that he was the Berlin otolaryngologist Aloys Maria Lautenschläger), as well as to a surgeon, Dr. Neupert, to discuss details and prospects of the implantation of the small disk. He envisaged patenting only the technical part of the invention and again requested Einstein's opinion.[210]

In August 1933 Goldschmidt wrote his next letter during a temporary visit in Paris. Einstein was staying in Belgium, preparing to leave for the United States. Goldschmidt was pondering on how to sustain his three sons, his daughter, and wife. "I have some new inventions," he wrote, "for which I miss your help . . . , especially in explaining phenomena."[211] He was also working on the hearing aid for Olga Eisner. He learned that the otosclerotics hear normally when on a railroad train, because of the stimulating effect of the periodic click, so he did experiments with Olga Eisner. They traveled in railroad cars with wooden (third class) and upholstered seats (second class); he made Olga Eisner travel on streetcars and built apparatuses that could produce equally effective stimulating noise, and so on and so on. Einstein replied to his letters rather quickly; but we know of them only because Goldschmidt mentioned them in his letters. In one, Einstein approved the stimulation by external noise; in another, he was curious how long the ability to hear returns after an impulse.[212] Otherwise, the only testimony that they ever met in person is Einstein's short letter to Olga Eisner stating that he met Goldschmidt in London and considers Goldschmidt's "new approach is the right one."[213] In 1933 Ein stein visited England several times from Belgium, so this encounter must have happened before October when he left Europe, and Goldschmidt with it, forever.

Goldschmidt stayed in England and did not give up his hope of developing the hearing aid to the point where he could take out a patent on it. "I hope that our happy collaboration may be renewed in spite of the distance," he wrote to Einstein in October 1941.[214] He had submitted a patent application a few days earlier on a simple device, consisting of an iron plate as a kind of diaphragm, fixed to the mastoid bone under the skin and an electromagnet applied to it outside. Goldschmidt was desperate enough to use himself as a test person. He had the plate affixed to his own skull, and "it was quite a sensation when I found that I can now 'hear magnetically' with my outer ear channels stopped up." He attached the description of the patent.

Even though Einstein, at the time already in Princeton, was happy to have a sign of life from Goldschmidt, he damped Goldschmidt's spirits by pointing out that "otherwise there are apparatuses commercially available here [in the United States] which . . . are as old as our old efforts. . . . Olga Eisner also has such an apparatus ('Sonotone')" and with this device there is no need for an operation.[215] Goldschmidt, however, did not withdraw his application, and a British patent on the device was granted to him in 1943. True, it was not shared with Einstein.[216]

In answering Otto Nathan's query whether Hans R. Goldschmidt-Goldie, Goldschmidt's son, had letters from Einstein,[217] he mentioned "the development of the bone-conduction hearing-aid (the Einstein-Goldschmidt patent)."[218] It is clear that the Goldschmidt patent shared with Einstein is *not* the hearing aid, but the magnetostrictive sound reproduction. From a psychological point of view, it is interesting to see how a patent with Einstein's name on it and a great story of how Einstein helped a famous singer were pasted together.

Another statement must also be corrected. Bruno Eisner maintained that the full development of the hearing aid was hindered by Goldschmidt's death.[219] As we have just seen, it was not hindered at all but was developed and patented.

Goldschmidt continued his correspondence with Einstein up to his death in 1950, but he never returned to the hearing aid.

When the Nazi secret state police, the Gestapo, visited Goldschmidt before he left Germany and searched for proceeds that he and Einstein might have derived from their shared patent, they found nothing.[220] Whatever Einstein contributed to Goldschmidt's inventions, he did it to help others—and for his own entertainment.

Goldschmidt, however, asked Einstein time and again to help him either in getting funds for his research,[221] in finding benefactors,[222] or in paving the way for patenting his inventions in the United States.[223]

CHAPTER FIVE

American Inventions

Altimeter

The following inventions were developed in cooperation with Gustav Bucky, a medical doctor, a specialist in X-ray examinations, and an inventor both in his field and out of it.[1] During his lifetime, he took out 143 patents. He is most famous for "Grenz ray therapy" of the skin. The rays of a 2-angstrom wavelength lay on the border (in German, *Grenze*) of ultraviolet light and soft X-rays; therefore, they are absorbed by a thin (3 millimeters) layer of tissue, and the irradiation of the skin does not affect the deeper layers of the body. His other invention still widely in use today is the Bucky diaphragm. It is a grid of lead strips that does not allow X-rays scattered at random angles by the irradiated body to reach the film, which would lower the quality of the image. When in 1933 Einstein met Gustav Bucky, his old acquaintance from Germany, in the United States, a friendship started that gave birth to a wide range of technological ideas.

Sometime in November 1934, Bucky proposed to Einstein that they construct a simple altimeter: a mass on a spring. When taken to a greater height, its weight diminishes, diminishing as well the elongation of the spring on which it hangs. An indicator will show this change. Other details are not known.

Einstein was not satisfied with the proposal.[2] By supposing that the mass lengthens the spring by 100 centimeters (40 inches), a change of 0.00001 of the gravitational field (the equivalent of an ascent of about 300 meters) would change the length of the spring by only 0.01 millimeters (0.0004 inches), which is hopeless to indicate. He came up, however, with another idea: to measure height by measuring the change of electric resistance of coal or graphite powder enclosed in a vessel in a vacuum. The powder is very fine, behaves almost like a liquid, and so it shows a definite electric resistance in a given gravitational field. The powder's density is highly sensitive to the change of the gravitational field, and this causes a change in its electric resistance that can be measured by a Wheatstone bridge. He pointed out, however, that the arrangement has to be protected from shocks. As a remedy, he proposed Cardan suspension. Furthermore, to avoid inhomogeneities in the powder, the electrodes should be built into the vessel's wall (fig. 5.1).

In a few days, however, Einstein came up with another arrangement.[3] Let us have an indicator in a vertical position with a spiral spring and a mass on the indicator (fig. 5.2). When we turn the indicator from its vertical position, the weight of the mass will help us, but the spring will counteract it. If we apply a third force to keep the indicator in a position tilted at a small angle, which can be the force of a weak spring or an electromagnetic effect, the arrangement will be sensitive enough to record variations of gravity with height. Tilting the whole arrangement would do the same. With two indicators tilted on the opposite sides of the vertical, the unavoidable oscillations can be counterbalanced.

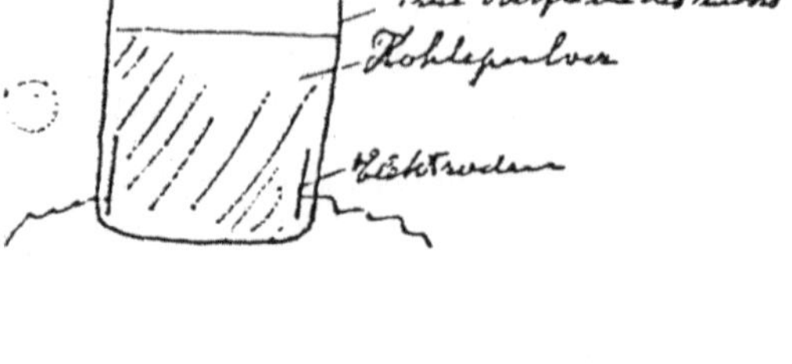

Figure 5.1. Einstein's first altimeter. Vakuum-Gefäß = vacuum tank; Freie Oberfläche des Pulvers = free surface of the coal powder; Kohlepulver = coal powder; Elektroden = electrodes.

Einstein to Gustav Bucky [November 8, 1934]. Courtesy Albert Einstein Archives, The Hebrew University of Jerusalem.

Figure 5.2. Einstein's second altimeter. Schwere = gravitation; Gewicht = weight; Spiralfeder = volute spring; Lager = bearing.

Einstein to Gustav Bucky, November 12, 1934. Courtesy Albert Einstein Archives, The Hebrew University of Jerusalem.

Einstein considered it would take years until a technically acceptable apparatus could be built—apparently an experience from his years working on the gyrocompass.

The altimeter was the first awkward step in their cooperation. Soon they launched a far more serious program: to set up a corporation for funding their research and testing their inventions.

On January 25, 1935, Bucky reported to Einstein that he turned to Emil Mayer, a technological counselor with broad experience and connections in the field, for advice.[4] Mayer proposed starting with a small mechanical shop, a mechanic, and a consultant physicist and engineer hired when needed. This would cost only a few thousand dollars. Under these conditions Mayer was ready to join them for a certain share. Then he proposed to bring up a capital of $50,000. For Mayer, Einstein's name was the main attraction to the venture.

Bucky set up a list of ideas.[5] Later someone (perhaps Helen Dukas, Einstein's secretary) dated it "1935." Some of the twenty-six items are related to Bucky's specialty, X-ray diagnosis and therapy: dosimeters, molding cones for irradiation, automatic irrigator, and cooler and heater pipe coils for film developing and fixing. Some are, however, surprisingly far from it: seats for vehicles; level measurement in car tanks; thermal bottle from paraffinated cardboard; car window dryer; springs with uniform pressure under load (e.g., for beds); non-moistening, gas-permeable fabric; automatic and mechanical telephone call counter; visual and acoustic signal for car speed control; bottle holder from pressed metal for more aesthetic serving; fastener for carving and serving knives and forks to prevent their falling into the pan. One item particularly strikes our eyes: an artificial horizon for aircraft based on a gyroscope.

Bucky was of the opinion that it would be wiser to start with a capital of $1,000 to $3,000 until one or two products were made.[6]

Einstein shared Bucky's opinion and even proposed to limit their activities to patenting, while leaving experiments and model construction to enterprises that could be approached with Mayer's help. He also agreed with carrying the financial burden of patenting by the three of them: Bucky, Einstein, and Mayer.[7]

Apparently in an earlier talk, Bucky and Mayer had agreed to submit the first group of inventions for patenting: the seat for vehicles, the nonmoistening fabric, the inexpensive thermal package, the liquid purification with strong electric field, and the gyro apparatus. Bucky even added that he had already found a workshop.[8]

For only some of these proposals can Einstein's share be established.

Waterproof Breathable Clothes

This special fabric would be made of threads of cellophane or other synthetic material, or of preimpregnated ordinary thread, and woven so tightly that the interstices are of the size of water drops (fig. 5.3). Capillary pressure would prevent drops from penetrating from the outside to the inside, but vapor particles (perspiration) and gases, having a smaller size, can leave from the inside to the outside.[9]

Before submitting the patent application, Mayer asked for further details, namely the ratio of the dimension of interstices to the thickness of the fabric or numerical values for threads of various thicknesses.[10] After having discussed the request with Einstein, Bucky answered that even 0.1-millimeter net size is enough to prevent water from penetrating the fabric under 1 millimeter of water pressure if the threads are nonmoistening.[11] They reduced the patent claims to garments and proposed to change the invention's name from "gas penetrable nonmoistening fabric" to "gas penetrable, watertight garment." Mayer accepted the proposal, translated the name into English as "improvements in waterproof fabrics" (Bucky, Einstein, and Mayer corresponded in German), and noted that the patent would be submitted in a few days.[12]

The copy of the application shows that Einstein and Bucky had been again asked to submit details, so they added maximum sizes of interstices in the fabric for various sizes of water drops (e.g., depending on whether the rain falls in a colder or warmer climate) and how double layers used on some parts of the cloth can prevent accidental moistening "without departing from the spirit of the invention."[13] The eight claims of the patent all begin with the phrase "Waterproof clothing, consisting substantially of . . . threads." Had they dropped "threads" and mentioned only "a fabric" with interstices of the size that they gave in the claims, they would have preempted by almost forty years Wilbert and Robert Gore and Rowena Taylor in patenting Gore-Tex.

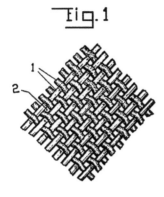

Figure 5.3. Waterproof garment.
Attachment to Emil Mayer's letter
to Gustav Bucky, June 5, 1935.
Courtesy Albert Einstein Archives,
The Hebrew University of Jerusalem.

INVENTORS
ALBERT EINSTEIN AND
GUSTAV BUCKY
BY *A. J. Kpenheiner*
ATTORNEY

Bucky forwarded Mayer's letter to Einstein with the copies of patent applications, but he was not happy.[14] He complained that Mayer had submitted the patent even though they (Bucky and Einstein) were against it. For what reason, I do not know. Apparently they succeeded in revoking the application, for there is no mention of a "Bu-Tex" or an "Eins-Tex" in their correspondence, and no patent was later granted for it.

Heat-Insulating Vessel

The application for another invention was submitted on June 5, 1935.[15] The walls of a cheap insulating package or vessel were to be made of two paraffinated cardboard sheets with glass fiber between them. The fiber isolates as a glass and, with its micropores, prevents air circulation in the wall.[16]

Next April, the patent attorney, Josef Oppenheimer, informed Bucky and Einstein that the patent examiner had reached the conclusion that "applicants have merely associated a plurality of desirable features found in the art, without producing a patentably new combination."[17]

Liquid Filtering by Electrostatic Method

The next common invention on Bucky's list to Mayer proposed a means of how to clean fluids that are in permanent use, for example, lubricating oils.[18] One side of a metal plate is galvanically coated with another metal, then cut into small pieces or even ground up. Contact between the two metals produces a strong electric field that attracts impurities and attaches them to the metal particles. The particles can then be washed and used repeatedly.

Before submitting the application, Mayer asked Bucky to mention a concrete example for the metals to be used.[19] After having discussed it with Einstein, Bucky answered Mayer's question the next day;[20] apparently the invention was Einstein's. Metals with a large enough difference in their contact potential, for example, zinc and copper, can be used. It is not necessary to bond the two metals firmly. It is enough to produce a conducting contact between the particles of their mixture by pressure. By July the patent proposals reached the Briesen & Schrenk office and, through it, the U.S. Patent Office. The examiner found three patents on the same invention.[21] In particular, the fundamental idea of Cabrera's patent (filtering with finely divided particles of zinc and copper and other metals)[22] looked the same as Einstein's, even though the two metals were not bound together in it.[23]

Einstein acknowledged that in a sense his idea had indeed been anticipated.[24] He found, however, a "misprint" in Cabrera's patent. When a zinc and a copper particle are suspended in a badly conducting fluid close to each other, there will be a potential difference between them but no electric field. For a field, the particles should be connected conductively. But to have two particles connected this way, they must be pressed together with considerable pressure. For this reason, Cabrera's filter cannot work.

Automatic Correction of Measured Data

This invention was first mentioned at the end of November 1934. Patent attorney Walter S. Bleistein, who was born in Berlin and studied engineering at Heidelberg, informed Bucky that he had attached a draft of the patent application to his letter, which was to be submitted in Germany.[25] In an undated document but apparently prepared before Bleistein's letter, Bucky and Einstein had formulated an agreement with Roefinag Research Corporation that they, as inventor partners, give the company the right and duty to apply for a patent in Germany for the same device and to promote its exploitation for a share in its profit.[26] My guess is Roefinag had three, and only three trustees: Bucky, Einstein, and Mayer.

The draft is of a device for automatic correction of measured data.[27]

The indication of measuring instruments is relative, the draft explains, because it depends on environmental conditions, such as air pressure and temperature. For precision measurements, corrections must be taken from tables or by calculations; or the correction factors are to be measured with another instrument; or multiple scales with already corrected values have to be used. The instruments can be divided into two categories: instruments that show a difference between the actual condition and a reference point; and instruments where the value to be measured is composed of two conditions: one that acts on the measuring instrument and another that can be arbitrary. An example of the first category is temperature measurement, where one of the thermoelements is located in the region whose temperature is to be measured, and the other thermoelement is kept at room temperature, which can change in any way. The measurement of humidity is an example of the second category. Here temperature influences the measured values.

Bucky and Einstein's invention is simple. The scale of the measuring instrument is shifted by a device that considers the condition for which the correction has to be taken. This way the pointer will show the corrected value. The scale can be calculated or empirically calibrated.

There are several variants of the arrangement. In the case of temperature measurement with thermoelements, the scale of the millivoltmeter is connected to a rod that dilates or contracts depending on the room temperature and moves the scale to and fro.

For measuring relative humidity, the scale can be moved by a device that is sensitive to absolute humidity, for example, by a hair; the scale is calibrated in relative humidity.

For pressure correction, the scale is moved by the pressure through a piston in a cylinder, which is connected to the source of pressure.

No patent can be found on the invention.

Airplane Horizon Indicator

The last invention mentioned in the list to Mayer was a "gyro apparatus" to be used as an artificial horizon or gyrocompass on aircraft. It was first described by Einstein in January 1935.[28]

The gyroscope is mounted on a peak or hangs on a thread. The gyro itself is driven not directly but by its friction with its rotated housing through the pivot and the surrounding air.

If the point of support of the gyroscope lies higher than the center of mass of the apparatus, the gyro works as an artificial horizon. The housing can be tilted without a major loss of independent motion of the gyro.

The housing can be suspended on gimbals. In this case, its rotation has a stabilizing effect, and the apparatus can also be used as a gyrocompass. Because there is no need for a special drive for the gyroscope, a relatively low speed of rotation can be used.

In a search for novelty in the U.S. Patent Office,[29] a patent by John Morrison was found,[30] which looked the same as this proposal. Einstein acknowledged that it completely anticipated it, and he was surprised that such an old patent existed without any practical application.[31] To display the instrument data, Morrison used a mirror on the gyro to reflect the celestial bodies. (This suggests that he intended to use it with a classical sextant.) He also proposed a special device to make the gyro rotate fast. This also suggests that the instrument was meant not for continuous but for occasional observation, whereas Einstein had in mind a continuous operation. This is an important difference between them. To make it clear, he complemented his original description with an optical arrangement (fig. 5.4). A picture of L can be seen on M through P, G, K, G. L, P, M, and the rotational axis of H are fixed to the airplane. The change in the angle of K entails a displacement of L's picture on M.

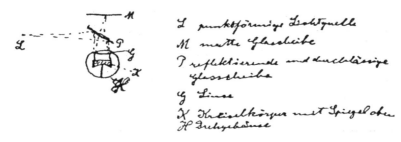

Figure 5.4. Artificial horizon. punkförmige Lichtquelle = point light source; matte Glasscheibe = opaque glass plate; reflektierende und durchlässige Glasscheibe = reflecting and transparent glass plate; Linse = lense; Kreiselkörper mit Spiegel oben = gyro with a mirror on its top; Drehgehäuse = turn case. Einstein to Briesen & Schrenk, July 13, 1935. Courtesy Albert Einstein Archives, The Hebrew University of Jerusalem.

Einstein then asked the question whether, with this amendment, the invention looked patentable. He did not get an answer, perhaps because Morrison had formulated the means of reading the output cautiously. He not only mentioned the observation of celestial bodies through a mirror but added the option to "fit other suitable optical arrangements to the gyrostat," and Einstein's new proposal was as such a suitable optical arrangement. Be that as it may, Einstein postponed further steps until a partnership with an interested concern could be found. "The evaluation of a patent by ourselves could defeat even here our situation, which is weak in every respect. Without capital and supplies, one is simply kicked out of the game, especially when one does not want to invest too much energy in it."[32]

On October 11, Bucky received a detailed description of the indicator from Einstein. Bucky retyped it and translated it into English.[33]

The text looks more like a plan than a patent application. The arrangement is fundamentally the same as described earlier, but now the optical attachment, the indicator, was added. Einstein considered the drive of the gyro to be the main novelty of the invention. This is a surprise because we have already learned that Morrison's patent used the same indirect drive. A real addition to the earlier description was the means of damping the precession of the gyro when it does not start its revolution horizontally. This is the air enclosed in the housing, which, when tilted, cannot rotate together with the gyro. It will, however, influence the rotation of the housing itself. This disturbance could be minimized by optimally adjusting the gap between the gyro and the housing and the form of the space the air fills. Damping can also be achieved by using a pivot with a spherical end instead of a needle-like one. If the frame is not fastened to the airplane but is suspended by gimbals, it will keep its horizontal position, and the direction pointing from the actual position of the light point on the screen to its position when the gyro is in a horizontal position will indicate the north-south direction. This compass function can be improved by putting an automatic connection between the frame and the plane that is to be kept horizontal, which, controlled by the light point, keeps the axes of the frame and of the gyro parallel.

In November 1940 Bucky requested a patent attorney, Walter S. Bleston, to look for companies who specialized in such devices. (Mr. Bleston's first name and middle initial were the same as that of Bleistein, their consultant in 1936. I am pretty sure they are the same person: Bleistein, Americanized as Bleston.) Bleston requested further details from Einstein.[34] After having summarized the gist of the invention, Bleston turned his attention to the fact that there were many gyros used as an artificial horizon that are driven not by electric motors but by an air turbine. With this drive they can reach their operational rotational speed faster than with Einstein's electric motor, and this is more useful for airplanes, which need a speedy indication. With dirigibles, faster indication is also welcome, because the larger mass of their gyros has a greater inertia.

Another problem can be the lack of lubrication. Finally, it is doubtful whether Einstein's gyro can keep its position on the pivot during the great tilts and jolts of the airplane. The air turbine gyros mentioned above are mounted not on a pivot but on the usual rings (gimbals).

With his questions, Bleston's intention was to be prepared for negotiations with companies such as Pioneer or Sperry, Anschütz's "enemy" in Einstein's first patent case. Apparently Bleston had got satisfactory answers from Einstein. At least we may guess that it was Bleston who arranged for a meeting between Einstein and the representatives of the Patent Department of Sperry Gyroscope Company on April 16, 1941, where he presented his invention. Surprisingly, he now replaced the electric motor of the housing with an air drive. The company, after having given "careful consideration to the air drive top-like gyroscope that you disclosed to us," gave a polite but reserved answer: "While superficially your design appears simpler than present designs, . . . it is believed that your design would prove more difficult to build and more expensive than the present model." The company considered it better to use gimbals and suspend the gyro pendulously. It promised, however, to build an experimental model to test, and to keep Einstein informed.[35]

In an undated letter, Einstein expressed his disappointment with Sperry. "They prefer selling something dear than cheap," and asked Bucky not to bother about "the thing." "If the thing is not made, the world will not end sooner than otherwise."[36]

The last words on this invention were written in a letter of 1942 (probably September 21): "The gyro man was not yet there."[37]

This was the last invention presented to Mayer as promising in early 1935, but it did not put an end to Einstein's cooperation with Bucky.

Electrostatic Microphone

"I am still waiting for the 'microphone,'" wrote Bucky in June 1935. "Are you no more convinced of it?"[38] More than a month later, Briesen & Schrenk sent an opinion about the patentability of an electrostatic microphone to Einstein.[39] Even though the company also sent a copy to Bucky, it did not attach the copies of patents mentioned in it. So it is highly probable that Einstein was its only inventor.

The idea of this microphone is that the sound waves compress or expand the air between the fixed plates of a condenser. The dielectric constant of the air changes with its pressure; consequently the capacity of the condenser will also change. The fluctuations of the electric potential give rise to fluctuations in the electric current. These are amplified by a back-coupling before they reach the amplifier.

This microphone is a variant of condenser microphones: they all make use of the changes of the capacity of a condenser either by changing the distance between the condenser's plates by using a diaphragm for the plate or by changing the dielectric constant.

Briesen & Schrenk found three patents that covered all or part of the claims of the Einstein microphone: Petersen's,[40] Speed's,[41] and Roberts and Ferriter's inventions. (I have not been able to find the last of these.)

In his answer,[42] Einstein proved by calculation that the Petersen patent (also a condenser microphone with no moving parts) cannot work, because the potential fluctuations give rise to current fluctuations that are smaller than the irregular fluctuations in the first electronic tube due to the atomistic constitution of electricity. This is the so-called Schottky effect. (The term used by American patents and correspondents was *shot effect*, for the irregularly arriving electrons shoot the anode of the electronic tube, giving rise to random noise. This is the "official" explanation. For me "shot" is "Schottky" with disappeared ending.) For this reason, Einstein continued, he made use of a back-coupling to have higher potential at the first cathode. It would, however, be best to test his idea and build an experimental device. Briesen & Schrenk agreed to this proposal.[43]

They looked for a company to finance the tests and approached David Sarnoff, president of the Radio Corporation of America. (Einstein first met him in April 1921 when he visited at the New Brunswick radio station.)[44] "I am a little bit anxious that Mr. Sarnoff will use our incomplete protection [by patent] to take the thing from us," Einstein wrote to Bucky. "But now we can't do anything more."[45] "We are completely dependent on the honesty and goodwill of the Radio Corporation," he emphasized in his next letter. "If it simply *wants* to rob us, we are powerless. . . . This is why I think that we should not handle the patenting ourselves but leave it on the Sarnoff people. This is why I think we should do nothing but wait for Sarnoff's letter. If he offers a connection with a decent fixed payment, perhaps I will accept it, and I will give you half of it. . . . If, however, the offer is stingy, then we drop the whole business and spare effort, work and cost thereby. . . . Don't be angry with me: for me, no money and no trouble is better than much money and much trouble. Nothing else but thinking and real work do I enjoy."[46]

In his much awaited letter, Sarnoff promised Einstein to send Max C. Batsel to discuss the microphone and the "precision measuring instrument."[47] Batsel was an engineer at the Engineering Department of RCA Manufacturing Company in Camden, New Jersey, which was to build experimental models and test them. "I am very anxious to establish a relationship between you and my company, and, of course, wish to do so in a manner which will be entirely agreeable to you." When Einstein gave a short account of this letter to Bucky, he noted that Sarnoff did not mention inventions, only Einstein's employment as technological adviser.[48] I do not know why he kept silent about Sarnoff's promise to build and check experimental models.

After having visited Einstein on July 25, Batsel requested an opinion on Einstein's microphone from his colleague, Ellsworth D. Cook, and forwarded Cook's eight-page opinion to Einstein.[49]

Even though Cook admitted that the linearity of the relationship between pressure and capacity recommends this idea for consideration, and the lack of moving parts is a further benefit, air pressures used in sound recording cause extremely small changes in electrostatic capacity.[50] Contemporary microphones can record sounds of the order of 0.05 dynes per square centimeter. Thermal agitation noise sets a lower limit, but the shot effect in the plate circuit of the first tube of the amplifier "appears to most definitely preclude the possibility of a practical 'dielectric constant' microphone."

Using realistic assumptions, Cook calculated the thermal agitation voltage for the two types of microphones and found the dielectric constant microphone too noisy for sound pickup and recording.

Almost at the same time Briesen & Schrenk had asked for a sketch of the microphone and details of the back-coupling,[51] but Einstein answered first to Batsel.[52]

He accepted Cook's results, but made two remarks. First, that by using an adequately high resistance in the scheme, the discharge current can practically be neglected with all the sound frequencies that are to be counted for. Thermal fluctuations of the potential are also eliminated by it. Second, the shot effect can be diminished to a harmless level by the back-coupling, already mentioned to Batsel. He also added that the first amplifier tube should be a low-voltage tube to avoid the electrons that hit the cathode and induce positive ions, which would destroy the grid.

He offered a more detailed written explanation.

The next day he answered Briesen & Schrenk and asked the company to shelve the matter: "Recently, mathematical reflections have convinced me that the microphone cannot work satisfactorily in the form you described."[53]

He informed Batsel, too, of his new calculations.[54] He accepted the importance of the shot effect. He had been convinced that the back-coupling would eliminate it, but new calculations convinced him that it was a mistake: the back-coupling would amplify this effect, just as it amplifies the primary current coming from the air condenser. The back-coupling's only benefit is to simplify the amplifier.

Then he compared the principles of electrodynamic and electrostatic microphones with an approximate calculation, with the result that the electrostatic microphone produces one-tenth of the potential the electrodynamic one does. This can easily be counterbalanced by a cascade, which can yield a potential ten to fifty times higher. In figure 5.5, 1, 2, . . . , 6 are isolated metal plates with alternating positive and negative potential P and $-P$. The air can move freely between them, with the exception of hatched gaps,

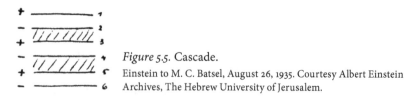

Figure 5.5. Cascade.
Einstein to M. C. Batsel, August 26, 1935. Courtesy Albert Einstein
Archives, The Hebrew University of Jerusalem.

which are isolated at their ends to keep air pressure in them constant. The potential fluctuations these condensers produce will add up, so we can have the same effect as with a single condenser but without impractically high potentials. The potential between the plates is maintained by very high resistance connections. This way it should be easy to produce 10,000 volts, and so to have a grid potential substantially higher than in the electrodynamic case.

Batsel politely acknowledged Einstein's comparison but considered it desirable to present "additional information in regard to the practical requirements as these have been established by experience with commercial microphones."[55] To obtain sensitivity acceptable for a commercial product, the electrodynamic microphones are heavier and larger than desirable, but with development their size can be reduced together with an increase of their sensitivity. Their impedance is quite low, and because the output of the microphone is connected between the grid and the cathode of a "high vacuum tube," which has a very high impedance, it is possible to step up their voltage by a transformer about 1,000 times. This is 10,000 times higher than what Einstein calculated for the condenser microphone, "a tremendous advantage insofar as noise due to 'shot effect' is concerned," and "even by means of the ingenious suggestion you have made for the cascade connection, the condenser type of microphone utilizing a variation in the dielectric constants of the air would not be practical." Their experience with condenser microphones that utilize diaphragms showed that it is difficult to maintain adequate insulation, and with the proposed type it would be even more difficult to cope with the higher insulation resistance.

Einstein was not convinced that a transformer could be used for raising the grid potential.[56] He remarked, among other things, that the self-induction of the primary coil cannot be very high; otherwise the higher frequencies would be cut off. The calculations being not easy, he proposed experiments.

In three days he realized that the high frequencies would not be cut off by the transformer. He proposed only making the inductive reactance of the primary coil large enough to make electrodynamic reaction low. "Now I am no more in a position to judge whether the dynamic or static principle is preferable. Both should certainly be tested carefully."[57]

In his answer, Batsel attached graphs and even a photo of an electrodynamic microphone with a transformer that "practically all of the broadcasting stations now use."[58] He concluded with the polite but rhetorical request to be advised on how to obtain a voltage from the static microphone that can compete with the electrodynamic microphone cum transformer. The same tension can be felt between them as earlier with Schuler in Kiel—in both cases a tension between the specialist engineer in the production line and the outsider scientist.

In a handwritten draft, Einstein doubted that a transformation of 1,000 is possible.[59] To ensure a fairly good transformation of lower frequencies, at least 300 windings are needed for the primary coil. On the other hand, it is practically impossible to have a secondary coil with more than 10,000 windings. This would give a thirty-fold transformation. He even questioned whether RCA had ever checked that a gain in intensity can be achieved by transformation without loosing proportionality. Apparently considering these objections too arrogant, in the typed letter sent he only asked for the actual number of windings in the primary and secondary coils and the value of magnetic permeability of the material the electromagnet's core is made of.[60]

Batsel sent him the requested data from the factory's design information used to produce this microphone.[61] The number of windings were 14 for the primary coil and 11,200 for the secondary one. The permeability of the nickel and steel alloy used for the magnet's core was around 2,500 under operating conditions.

Einstein noticed that the inductive reactance is not much higher for low frequencies than the resistance given in Batsel's letter, even though earlier Batsel had emphasized that a large difference is needed to make sound reproduction uniform.[62] At these low frequencies, only one-twentieth of the induced primary potential can be multiplied by the transformation number of circa 800. Even though for higher frequencies this ratio turns better, the internal capacity of the secondary coil comes into play, and this is not so easy to calculate. "All in all it seems to me therefore that when using a transformer, the amplification must be paid for with a deterioration of the independence of the transmission from the frequency. Therefore the reason for using a transformer seems to me dubious."

Before answering Einstein's letter, Batsel again turned to Cook for help. Cook performed calculations for both 50 cycles and 10,000 cycles and reached the conclusion that "the response at both 50 and 10,000 cycles is as satisfactory as is commercially necessary."[63]

In addition to forwarding Cook's letter, Batsel politely reminded Einstein of data that had already been sent to him in his previous letters (and which were not meticulously followed by Einstein in his replies).[64] "I hope we have been able to make clear to you in this letter the values of the constants that have been used in these designs, as I am sure that we have not previously stated these facts in such a way that they were clear to you."

Meanwhile Einstein was waiting for Sarnoff to enter the ring and continue their discussion on the tests and especially on the "relationship" between Einstein and RCA mentioned in Sarnoff's letter. By the end of October, he lost his patience. Apparently reluctant to turn to Sarnoff himself, it was Elsa who took things in her hands.

"The business affair that was under consideration by you and my husband did not reach a fortunate conclusion at the time," she wrote (in English) to Sarnoff in October 29.[65] "I may say to you that this has been a matter of deep regret, both to my husband and to myself." She blamed circumstances, not Sarnoff, and asked whether the negotiations could be resumed. In addition, "it would be a great pleasure to him [Einstein] to be permitted to act as scientific consultant for your Corporation, and he would, perhaps, accept this under definite terms that both of you could agree upon between you. My husband has such a great joy in solving technical problems and in Germany he formerly worked in this field from time to time." It is remarkable that the terminally ill Elsa Einstein took on the awkward task of begging for a position for her husband. Awkward all the more so, because she got no answer.

A half year later, on May 20, 1936, in a letter written in English, Einstein reminded Sarnoff, that "almost a year has elapsed since our first conference. Since that time my relations with Dr. Bucky have become so diminished with regard to technical matters that I would be able to enter into closer association with a technical enterprise without him and with no hard feelings whatsoever. Therefore, I should be glad if you would again revert to your proposal made on that occasion."[66]

He brought up two reasons why: his "ever keen interest in technical problems," and his hope "to earn the means to make possible to a couple of worthy young colleagues continued employment on theoretical problems in these times that are so difficult economically." He would not accept a fee but proposed that RCA pay "the proper compensation, suitably increased from time to time," to these young scientists. The arrangement will also help his own work, for he would not be forced to seek for assistants. At the end of his letter, it turns out that he had in mind his young assistant, "a Russian Jew, born in America," whom he would lose without such a help. The young colleague was going to accept a position at the University of Rostov. To mention the assistant's origin was possibly meant to explain how an American could accept a position in Russia, but might elicit Sarnoff's sympathy, because Sarnoff was a Jew born near Minsk.

Sarnoff promised to give "careful consideration" to Einstein's suggestion and promised an early reply.[67] On June 7, apparently upon the young assistant's inquiry, Einstein could not say anything reassuring, only that "Sarnoff has still not answered!"[68] His letter to the young assistant is addressed to Nathan Rosen, the coauthor of two papers that are considered to be among the best works Einstein published in the United States: one presented what became known as the Einstein-Podolsky-Rosen paradox, whereas the other described a wormhole between a black and a white hole, now called an Einstein-Rosen bridge. In view of his achievements and talent, Rosen deserved the attention Einstein paid to him, and Einstein did his best to help him get a decent position, if not in the United States, then in the Soviet Union. He even sent a letter of recommendation to the chairman of the Council of People's Commissars (prime minister), Vjacheslav Molotov, in March.[69]

At the end of July, a letter arrived from Engbert S. Reid, secretary to Sarnoff.[70] He addressed his letter to Helen Dukas. Apparently Einstein had requested that she, and not the terminally ill Elsa, remind Sarnoff of his promise. Reid confirmed that Sarnoff had not yet written the letter, but he would remind him to do so upon Sarnoff's return from his tour in Europe. The letter was never written.

With Molotov, Einstein was more successful. On July 4 he expressed his gratitude for Molotov's help.[71] Rosen moved to the University of Kiev the same year and worked there for two years before returning to the United States and eventually settling in Israel, and becoming president of Ben-Gurion University.

Fluid-Level Indicator

On June 27, 1934, Bucky submitted a patent application to the U.S. Patent Office for an oil tank level indicator. The patent was granted next February.[72]

For some reason or other, he was not satisfied with it. In June 1935 he reported to Einstein that he now had found a satisfactory solution for such a gauge, so important for cars.[73] He proposed one (*I* of fig. 5.6) or two funnels (*II* of fig. 5.6) to immerse in the fluid and illuminate their surfaces. The diameter of the visible circular surface is proportional to the height of the fluid. The second solution is a variant of the first: two circular surfaces change their diameter and the distance between them with fluid depth.

Another solution is to immerse a stick with little mirrors or little reflecting spheres on it at different distances. A third possibility is to wind a spiral of reflecting tape around the stick (*III* of fig. 5.6). When illuminated, the light reflected on a transparent screen would indicate where the level of the fluid is. These indicators are cheap to make. During the day, daylight can substitute for the electric light.

An undated letter seems to be Einstein's reply.[74] He was not happy with Bucky's proposals, because dirt in the fluid would influence observation. It

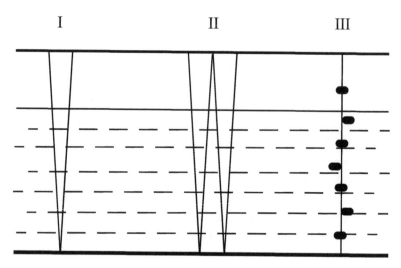

Figure 5.6. Bucky's fluid level indicators.
After Gustav Bucky to Einstein, June 7, 1935.

would be better to use the surface of the fluid itself for reflection, as Bucky had done in his first proposal.

Einstein came up with his own idea, too: to use a funnel—made of light, a light cone (fig. 5.7). The light source is projected on a semitransparent mirror. It is reflected on the surface of the fluid and back through the semitransparent mirror to an opaque screen. The lower the fluid level, the smaller the spot on the screen. He was, however, not satisfied with the whole idea. "I think there is no really good *optical* solution."

All the same, Bucky had prepared a description for patenting (which is not available). Apparently he proposed following the level of the fluid through the cap of the vessel by making it transparent. Einstein found three problems.[75] For one, the cap may get covered by fluid drops when outer temperature falls below inner temperature. Second, if the vessel is a gas tank, the driver can read the level in his driving seat only indirectly by means of an optical arrangement not specified in the description. Finally, the patent is formulated so peculiarly that it could easily be evaded. He then added: "I would prefer not to act as a coauthor with these two patents. Not only did I not take part in their development, but I cannot even believe in its practical implementation." (The other patent mentioned here is the self-adjusting camera, described in the next section.)

As Einstein repeatedly emphasized to Bucky, his main problem was how the driver would be able to follow the gas level from his seat, but now that Bucky had reassured him that it can be surmounted, his only headache left was to find "auto people" to finance the development.[76] Indeed, Bucky prepared a new description in the form of a patent application.[77] He presented not only his original three versions but also Einstein's light-cone version.

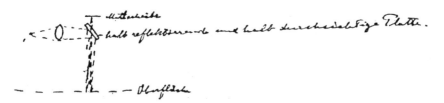

Figure 5.7. Einstein's fluid level indicator. Mattscheibe = opaque screen; halb reflektierende = semitransparent [mirror]; Oberfläche = [liquid] surface.
Einstein to Gustav Bucky, after June 7, 1935. Courtesy Albert Einstein Archives, The Hebrew University of Jerusalem.

He also specified the device by which the level can be read from the driver's seat.

Einstein came up with a new, electrical (not optical) idea of indicating level change with a conductor mounted in the tank vertically.[78] Let us warm it by an electric current. If the conductor is made of a material whose resistance depends on its temperature, the part submerged in the oil will be cooled faster than the part above it, because that is cooled only by the surrounding air. As the level of the fluid lowers, less and less is covered by the oil, so the conductor's average temperature will grow and so does its resistance. Level indication is reduced to resistance indication.

The resistance can, however, change with temperature outside of the tank, too. This disturbing effect can be eliminated by using two conductors of the same material and same cross section but of different form: a band and a wire. Because of its larger surface, the band conductor loses heat faster than the wire, while their rest resistances stay the same because of their equal cross sections, independent of outside temperature. Let us connect the two conductors in a Wheatstone bridge (fig. 5.8). Here *a* and *b* are resistances of equal size; the conductor on the left is the wire, and on the right it is the band. If we put a certain potential between *A* and *B*, the current measured by *C* will depend only on the oil level.

The band-form conductor can be replaced by a conductor that is in thermal contact with a mass of high thermal capacity to prevent its warming up. There are other possibilities, too.

In November 1936, Einstein disclosed one of those possibilities: to use thermoelements on a stick at different heights and connect them in series.[79]

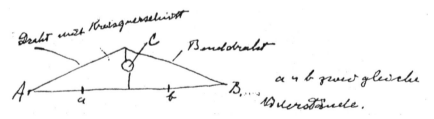

Figure 5.8. Wheatstone bridge. Draht mit Kreisquerschnitt = wire with circular cross section; Bonddraht = stripe; *a* u. *b* zwei gleiche Widerstände = *a* and *b* resistances of equal value.

Einstein to Gustav Bucky, no date. Courtesy Albert Einstein Archives, The Hebrew University of Jerusalem.

Let the stick be immersed in the oil tank, and let us add a thermoelement outside of it and in thermal contact with a conveniently high thermal capacity. If we send an alternating current through the circuit made of them, the thermoelement outside the tank will not warm up, and neither will those elements which are immersed in the oil. Those, however, that are out of the tank in the air will lack this cooling.

No patent was granted.

Light-Intensity Self-Adjusting Camera

An automatic camera was first mentioned in Einstein's letter of around July 9, 1935.[80] In it, he brought forward three difficulties: the photo current is too weak to move the screen that controls light intensity entering the camera; balancing the camera in a tilted position is difficult; and exposure time cannot be shortened by using high light intensity. It is easier to understand what he was writing about if we look at the figure in the final patent description.[81]

In *Fig. 1* of figure 5.9, *13* is the objective of camera *10* shown in side elevation; *15* is a commercial photoelectric cell with a mechanism to move screen *23* in the beam of incoming light, for example, between two of the objective's lenses. The screen is transparent and is shaded from clear to black (*Fig. 4*). Its function is to reduce incoming light continuously, controlled by the photoelectric cell. *Fig. 2* and *Fig. 3* show diaphragms to a rougher setting of light intensity for different shutter times; the automatic screen performs the fine tuning.

Einstein found a mistake in one of the figures.[82] The variable diaphragm (*Fig. 2* and *Fig. 3*) must be mounted where the light beam is narrowest, and at the objective and not close to the plate. The stripes of the screen (*Fig. 4*) should also be as wide as the opening of the diaphragm. He was not satisfied with the formulation of the main claim either. Because the final patent description shows that it had been amended according to this proposal, Einstein's remarks were made before the patent was granted, that is, before October 27, 1936, perhaps even before it was submitted on December 11, 1935.

The patent was granted for both of them. Einstein felt uncomfortable: "I am very happy that the movable automatic diaphragm has been patented," he wrote to Bucky. "I have, however, not given this thing any of my brain fat at all. I must really stress it."[83] There was, however, nothing to break the charm of his

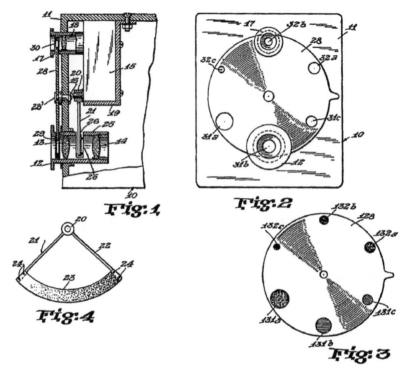

Figure 5.9. US2058562. Light Intensity Self-Adjusting Camera

name. When the word on the invention reached the press, it was Einstein and not Bucky who appeared in the headline of the *New York Times* as the "inventor of camera device," and Bucky was relegated to the subheading.[84]

Soon after patenting, the inventors submitted an application to the German Patent Office through Roefinag Research Corporation.[85]

Steel-Tape Recording

When reading Einstein's letter of March 9, 1936, to Bucky, we are surprised to learn that they intended to realize a "proportional recording of sound waves in a magnetic way,"[86] that is, to invent a tape recorder. Surprised, because magnetic sound recording on steel tape had already been invented and used since 1929 in Germany and the United Kingdom, even though not on a commercial scale. This invention, however, reached American coasts only in the late thirties, which may explain why Bucky considered his idea worth patenting.

The letter is a lone document, so we do not know about Bucky's proposal. Einstein apologized in it for his earlier objections and explained why now he considers the process physically realizable. "When a steel band is simply magnetized with a moving alternating field," he wrote, "the curve of magnetization certainly is disgustingly distorted," but it can be remedied in the following way.

Let us consider a piece of iron that is magnetized in direction R with a strong magnetic field. We will have a remanent magnetism I in this direction. If we use a weak magnetic field to magnetize the iron in the opposite direction, it will lower the magnetization by ΔI, which is proportional to the weak field. If we use an alternating weak field to magnetize a previously magnetized steel strip, only the half waves will be recorded that have the direction opposite to the original magnetization R.

Now let us choose as a magnetizing field the superposition of a constant magnetizing field H (Einstein wrote, apparently by mistake, "demagnetizing field") and a weak field h to be recorded and keep the amplitude of h always lower than the amplitude of H. Because the field is $H + h$, and H is constant, we will have a recording that is exactly proportional to the weak field. "This should be realizable, and it will surely be superior to the gramophone disc," he concluded.

If Bucky had succeeded in getting a patent, he and Einstein would have been commemorated as pioneers of tape recording in the United States.

Aircraft Speedometer

"While riding on a train, a solution for a speedometer for airplanes occurred to me, which fascinated me so much that I consider it Columbus's egg," Bucky wrote to Einstein on September 8, 1942. And after having described the idea, he closed the letter with the question: "Does the egg smell?"[87] I am sure you recall that once upon a time Columbus solved the problem of how to stand an egg on its top by gently hitting it on the table and cracking its top a little bit. At least this is the popular explanation of this bon mot for a simple trick.

Bucky proposed shooting small balls with a mechanical device one after the other within the airplane perpendicular to its axis and exposed to the wind; after hitting their target, the balls roll back. The measure of the plane's velocity is the change in the trajectory of the balls. The trajectory is determined by the wind pressure, the velocity of the shot, and the acceleration [*sic*] due to the airplane. The first two can be measured independently, so with the help of the deviation of the trajectory, the third, the velocity of the plane, can be determined.

When the airplane does not move and we shoot a ball out of it perpendicular to its axis, he continued his confused considerations, the ball will follow a perpendicular straight line. If the wind blows, the velocity of the ball will be the vectorial sum of the velocity of the shot and the velocity of the wind and its orbit a straight line (!).

If the airplane flies in still air (?), the ball's velocity will be the vectorial sum of the plane's acceleration and the shot's velocity.

If, in addition, the wind blows, the trajectory will take the form of a parabola.

If we keep conditions constant, the ball will always hit the target at the same spot, Bucky continued. The effect of changes in wind velocity can also be eliminated by a simple mechanism. If we mount a series of electric contacts on the oblong target, the ball will hit one of them according to the airplane's speed, and the signal can be followed on a screen with illuminated stripes on it. "The technical side is clear to me and presents no insurmountable difficulties," he added.

I can imagine how wide Einstein's eyes were open when he read this confused explanation. What the pilot really sees is simply that the balls always fly on parabolas starting perpendicular to the plane's velocity, due to the constant pressure of the wind. With this arrangement the plane's velocity can be measured with respect to the surrounding air. To use the term *acceleration* for *velocity* of the plane is a venial sin for a medical doctor. Bucky's letter makes us understand why he needed a trained physicist to check his ideas.

Einstein did not want to tell Bucky how much the egg smelled.[88] He explained to him that during its flight the ball will hit the target at a spot at a distance from the spot it would hit if the airplane were at rest because of the constant pressure of the wind. With constant distance between "gun" and target, this shift depends only on the velocity of the plane *with respect to the air*, and with it the task would be solved. This means Einstein tacitly accepted that Bucky intended to measure airspeed and not ground speed. Or this was

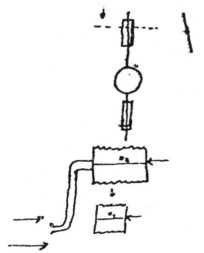

Figure 5.10. Einstein's air velocity meters.
Einstein to Gustav Bucky, October 8, 1942.
Courtesy Albert Einstein Archives, The Hebrew
University of Jerusalem.

the way he explained to him that what he wanted to invent was an airspeed indicator.

Air pressure depends, however, not only on the plane's velocity, Einstein continued, but also on the density of the air. To eliminate this effect, Einstein proposed two possibilities (fig. 5.10).

The first is a wind wheel with slightly tilted vanes. With resistances disregarded, its revolution is exactly proportional to the plane's airspeed. This device would then work as car speedometers do.

The second possibility is to measure velocity by measuring the air pressure "in the usual way." This means using a Pitot tube looking in the flight direction, which measures the total pressure in a chamber, and have another chamber with holes looking perpendicular to the flight direction to measure static pressure. The difference between the two pressures gives dynamic pressure, the pressure due only to flight velocity. In the figure, a tube enters both chambers. I do not know whether it is a mistake and the tube should enter only the upper chamber as a Pitot tube, or whether it is another arrangement that, for lack of details, I cannot interpret. The pressure measured in the chamber is x_2.

As already mentioned, air pressure is proportional not only to the speed of the air but also to its density. If, however, the density is compensated for by compressing the same air (in the figure it seems the same chamber is compressed), "the compensating pressure measured in volume ratio will also be proportional to the air density," Einstein continued. Let x_1 be the compressed pressure; then the ratio $\dfrac{x_2}{x_1}$ is independent of air density. The compression must, however, take place at ambient temperature, he added, which makes the realization of the idea difficult. Ice formation is a further problem, but it can be avoided by putting everything in a box.

Did Einstein invent the airspeed indicator of our days? There is no need to trace its history far into the past. An airspeed indicator was invented and patented in 1940 by Caltech graduate Leo N. Schwien.[89] It not only made use of a Pitot tube and another one for static pressure but also compensated for the changes of air density and its temperature.

Einstein's idea is original, but both his and Bucky's plans went nowhere because of their ignorance of the contemporary literature.

Timer

Probably in the fall of 1942, Einstein sketched a timer for Bucky (fig. 5.11).[90]

Let us have a closed, parallelepiped tank supported on two opposite sides at the middle. It is half-filled with water and has a vertical wall in it with holes at its top and bottom. Without the holes, the tank would be in equilibrium in any tilted position. Let us have an excess weight and a contact lever on the left.

When the electric current to be controlled is turned on, let the device set in equilibrium mechanically and be released. How long it will tip over depends on the size of the excess weight and of the lower hole on the wall. It can be set comfortably to 1 second.

It is possible that Bucky's letter to Einstein, written on September 19, 1942, is a reply to this proposal.[91] "Sure your gadget is very nice for stationary operation. We should, however, operate the camera in every possible posi-

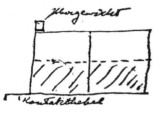

Figure 5.11. Timer. Übergewicht = excess weight; Kontakthebel = contact lever.

Einstein to Gustav Bucky [before September 19, 1942]. Courtesy Albert Einstein Archives, The Hebrew University of Jerusalem.

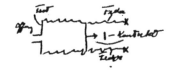

Figure 5.12. Timer versions. Offng = opening; Fest = fixed wall; Feder = spring; Kontakt = contact.

Einstein to Gustav Bucky, [September 21, 1942] Montag. Courtesy Albert Einstein Archives, The Hebrew University of Jerusalem.

tion. As far as I understand, the tip-over presupposes a horizontal position." From this we may guess that the timer was to be used with the automatic camera.

"I did not know that the camera had to be used in any position," answered Einstein.[92] The same idea of letting a liquid flow through a narrow hole can be applied even in this case. One must only use a vessel with elastic walls like an accordion (fig. 5.12).

Air can also be used in place of the liquid, for atmospheric air behaves like an incompressible fluid if its volume in the vessel is not too large. "If the thing [the "accordion"] were released when the lamp current is closed," he continued, "then adjusting the spring's force can easily determine the desired length of time before the contact closes. The springs should rather be set aslant so as not to be dependent on gravity and shocks."

We have no details of how this timer would be used.

On the Threshold of the Manhattan Project

In early December 1941 Vannevar Bush, chairman of the National Defense Research Committee, approached Einstein, through the director of the Institute for Advanced Study, Frank Aydelotte, with a theoretical problem.

We do not know the exact question, but it was connected with the enrichment of natural uranium in its lighter isotope, uranium-235, by gaseous diffusion. Uranium-235 was going to be used as the explosive for one of the future bombs. How much Aydelotte and Bush disclosed to Einstein regarding its importance for developing the bomb is unknown, but that the question was of military importance was made clear to him: "I know how deep is his [Einstein's] satisfaction at doing anything which might be useful in the national effort," Aydelotte wrote to Bush in his letter accompanying

Einstein's handwritten calculations.[93] (He was asked, for reasons of security, not to type up his letter.) Aydelotte remarked that Einstein offered to continue the calculations "if there are other angles of the problem that you want him to develop."

Bush forwarded Aydelotte's letter with Einstein's calculations, made in a couple of days, to Harold C. Urey, at the time professor and director of War Research, Atomic Bomb Project at Columbia University, with the remark that Einstein "treated only the classical problem."[94]

Meanwhile Aydelotte repeated Einstein's readiness to continue if the "direction" in which to continue would be indicated. He also forwarded Einstein's opinion that "some aspects of the problem could be solved . . . by a few simple experiments."[95]

Urey answered Bush on December 29.[96] "Since he [Einstein] has taken the mean velocity he has eliminated part of the problem," he wrote. "Perhaps there is a flow through the small pores of our membranes, but superimposed upon this is diffusion as well," and Einstein's calculation is too simple to tackle the problem. Was Einstein considering only the flow and neglecting diffusion?

From Bush's letter to Aydelotte, written the next day, we learn that Bush had wrestled with the same problem "some fifteen years ago, . . . and I got thoroughly stuck."[97] With his simplifying assumption, he continued, Einstein modified the problem, but in this form it had already been treated by others. Sure, with more information given, Einstein could continue, but "the reason that I am not going further is that I am not at all sure that if I place Einstein in entire contact with this subject he would not discuss it in a way that it should not be discussed, and with this doubt in my mind I do not feel that I ought to take him into confidence on the subject to the extent of showing just where this thing fits into the defense picture, and what the military aspects of the matter might be." He would be happy to give Einstein all the information he needs, but "this is utterly impossible in view of the attitude of people here in Washington who have studied into his whole history."[98] It is better to find another physicist for the task, because the problem with Einstein's person outweighs the expected benefit.

Urey's detailed answer to Bush underlined this opinion.[99] Einstein used "the most general type of mathematics," Urey wrote. "He has never done anything in which careful long difficult detailed calculations have been involved." (An interesting remark about the creator of general relativity and pioneer of the unified field theory.) The solution of the problem needs such hard work for weeks, he continued, with numerical results. Urey's conclusion was that "it is not possible to give Einstein this specific problem and expect results." Bush accepted this opinion and approached Norbert Wiener.[100]

We cannot get closer to the problem that Einstein was supposed to solve by looking into Bush's work done "some fifteen years ago." The only paper I have managed to find from Bush's pen on gaseous diffusion was published in 1922.[101] In it, Bush and Smith set themselves the goal to construct gaseous conduction devices to be used, for example, as rectifiers. They investigated the penetration of gaseous ions through porous walls, but in the paper they did not enter into details.

Einstein had to wait for two years before he was permitted "to be useful in the national effort."

Torpedoman Einstein

On May 13, 1943, Lieutenant Stephen Brunauer, in a follow-up of his earlier visit, turned to Einstein with the request to cooperate with the Research and Development Division of the Bureau of Ordnance of the Navy Department.[102] Having got the consent of his "boss," Frank Aydelotte,[103] Einstein was happy to cooperate and looked forward to fulfilling his tasks.[104]

Stephen Brunauer, born in Hungary as István Brunauer, left his home country because Hungarian universities limited the enrollment of Jewish students after the First World War. He attended the City College of New York and Columbia University, graduating magna cum laude in chemistry and in English. While working at the Fixed Nitrogen Laboratory of the Bureau of Chemistry and Soils of the United States Department of Agriculture under Paul H. Emmett, he obtained his Ph.D. at the Johns Hopkins University in just a year. Then, together with Emmett, he completed the experimental part of a project that later became known as the Brunauer-Emmett-Teller (B.E.T.) method after Brunauer's compatriot Edward Teller provided the theoretical basis for their experimental work. In 1939 the three scientists published the method with which one can calculate the surface area of finely divided powders from the amount of gas adsorbed on them.[105]

Even though Brunauer won a name for himself with the B.E.T. method, the laboratory wanted to focus on other research areas, so he had to look around for an appropriate position. Not long after Pearl Harbor, he was called up for active service, and as a lieutenant, junior grade, was charged to head the high-explosive research and development group in the Bureau of Ordnance of the Navy.[106]

Einstein's name cropped up at one of the joint meetings of the research authorities of the Army, Navy and the National Defense Research Council.[107] Brunauer, curious, asked whether Einstein worked for any of them. He got answers like "Oh, he's a pacifist," or "He's not interested in anything practical . . . only in working on his unified field theory." For some reason or other, Brunauer could not imagine that Einstein was not interested in a war with Hitler, so he wrote Einstein and asked for an appointment.

As a matter of fact, it is highly probable that Einstein and Brunauer were not unknown to each other. In 1931 Brunauer married Dr. Esther Caukin, a member of the American Association of University Women. Einstein, in his effort to find positions for academic women who were forced to leave Germany, turned to the association, and he kept in touch with Caukin-Brunauer at least from April to December 1938.[108]

Einstein's answer to Brunauer's (nonextant) letter sent in 1943 was positive, so they met sometime before May 13 in Einstein's home. (Even though Brunauer, in his short account, sets the date of their meeting at May 16, it is a mistake. Brunauer confirmed the result of their discussion in his letter of May 13.)[109]

Brunauer put the question straight to Einstein, whether he would become a consultant for the navy in the field of high explosives research. Einstein was happy to say yes, and Brunauer was happy to say "you are hired." "People think that I am interested in theory and not in anything practical," Einstein added. "This is not true. I worked in the patent office in Zurich, and I participated in the development of many inventions. The gyroscope too."[110]

Brunauer's success in recruiting Einstein for the Bureau of Ordnance created a sensation. "Einstein is one of us," they felt. Einstein must have been happy, too, if not for any other reason than for the fact that, as he told to Brunauer, "I am in the navy, but I was not required to get a Navy haircut."[111]

To be sure, he was not in the navy. He was to sign a research contract with the navy as a civilian. As Brunauer explained to him in his May 13 letter, the navy used two types of contracts for civilian services: one with an institution, and another with an individual. Brunauer added that Einstein might even work for them informally, without any contract. After consulting with Aydelotte, Einstein chose to work on contract as an individual. Brunauer offered a daily fee of $25, as the allowed maximum—"a ridiculously small fee," he remarked in his reminiscence. He also indicated that Dr. Kirkwood would introduce

Einstein to the work done so far. John Gamble Kirkwood was a professor at Cornell University, whose specialty was the study of shock. Brunauer added that he had asked E. Bright Wilson to contact Kirkwood. Wilson was research director of the Underwater Explosives Research Laboratory in Woods Hole, Massachusetts, an alumnus of Princeton University with a Ph.D. in physical chemistry from Caltech.[112]

Brunauer proposed May 30 for his visit together with Kirkwood[113] and informed Einstein that John E. Burchard, division chief of the National Defense Research Committee, had authorized him to use the bureau's library.[114]

There are no details about Kirkwood and Brunauer's visit with Einstein, or even whether they met at all. The next two letters, however, indicate that Einstein was given the task of coming up with an idea on the most effective explosion of torpedoes, as a practical use of the hydro- and aerodynamical studies performed by Kirkwood and others (among them, as we will see, John von Neumann).

Why just torpedoes?

The "Great Torpedo Scandal" had its peak between the end of 1941 and August 1943.[115] The main problems were depth control, the unreliability of the magnetic-influence exploder, and also of the contact or impact exploder. We are interested in the second and third of these problems. The contact exploder works when the torpedo strikes the target; the magnetic-influence exploder, or proximity fuser, detonates the warheads when under a warship. The problem with the contact exploder was that it did not work properly when the torpedo struck the target at 90 degrees; the problem with the proximity fuser, especially with the type Mk. 6, was that it sensed the variation of the horizontal component of the Earth's magnetic field as the torpedo approached the target ship. Because this component is inversely proportional to magnetic latitude, the fuser became more and more unreliable near the magnetic poles. Because of their failures, proximity fusers were abandoned at the end of 1943.

The navy's attention then turned to impact exploders, which also had a problem: the elevated speed of the torpedoes resulted in higher inertial forces at impact, which often bent the vertical pins that guided the firing pin block.

My guess is that Brunauer set Einstein the task of reconsidering magnetic-influence exploders, as will be clear in what follows.

Einstein turned to two of his colleagues, (Walter Porter?) White and a certain Taub, for help in digging up literature.[116] I know nothing about his success in getting adequately informed, but he came up with the following arrangement (fig. 5.13). *S* is an electromagnet that produces an alternating magnetic field; the two *s*'s are electric coils, connected in series and producing magnetic fields with opposing polarities when *S* induces current in them. The device is to be put in action when the torpedo has already moved far away from the attacking ship to avoid triggering by its magnetic field.

The torpedo must be shot deep enough to pass below the target. As it approaches the ship, the hull will induce an additional magnetic field in the two coils *s*, first stronger in the first, then stonger in the other, creating an electric current whose intensity will change according to the following curve (fig. 5.14). When the torpedo passes the middle of the ship, *M*, the fields from both sides of the hull would cancel each other, and the current would briefly dip to zero. What we need now is an electric device that functions when an initially growing current goes to zero. Einstein cautioned, however, that the idea was probably not new or perhaps technically difficult to realize.

The answer to this proposal came not from Brunauer but from the chief of the Bureau of Ordnance, Rear Admiral William H. P. Blandy, partly as a "welcome on board," partly as a courteous reprimand on how to label confidential letters.[117] "Your letter was correctly mailed in a double envelope with the word 'confidential' on the inner envelope," he wrote. "Beside this, the regula-

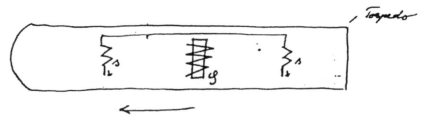

Figure 5.13. Torpedo blast.
Einstein to Stephen Brunauer, June 18, 1943. Courtesy Albert Einstein Archives, The Hebrew University of Jerusalem.

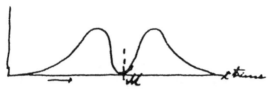

Figure 5.14. The time change of electric current.

Einstein to Stephen Brunauer, June 18, 1943. Courtesy Albert Einstein Archives, The Hebrew University of Jerusalem.

tions prescribe that the word 'confidential' be stamped clearly on every page of the letter."

Blandy, however, did not only admonish this civilian called Professor Einstein of his obligation to follow strict rules in his correspondence but reflected on Einstein's idea, too. He remarked that there are devices in the navy similar to Einstein's and offered to send two experts for further discussion: John Bardeen and Frank Brown. John Bardeen who, after the war, received two Nobel Prizes in physics, one for the transistor in 1956, and another one for a theory of superconductivity in 1972, worked for the Naval Ordnance Laboratory as a civilian physicist on how to protect U.S. ships and submarines from magnetic mines and torpedoes.[118]

Bardeen and Brown visited with Einstein on July 2, 1943.[119] A memo summarizes Einstein's proposals as follows: "Let a primary coil be energized with alternating current and let two secondary coils, located symmetrically forward and aft of the primary, be connected in opposition. The axes of the coils may be all vertical or all horizontal. Then no signal will be received from the ship or when the primary coil is directly under the keel, but as the torpedo passes under the side of the ship, the symmetry will be disturbed and a signal will be received. The circuit is to be arranged to fire when the alternating voltage returns to zero after passing through a maximum."

An identical exploder had already been tested at the Naval Ordnance Laboratory, where it had encountered engineering problems, including suitable frequency, attenuation and phase shift due to the sea water, eddy currents, and coil geometries. "Dr. Einstein's suggestion relates primarily not to the engineering aspects of the problem, but to the problem of localizing the firing close to the keel. . . . This is an important part of the general problem, but a detailed consideration of it has had to be postponed because of the pressure of other work," the memo concludes. A further problem is the localization of firing, especially when the target ship has a flat bottom. Einstein also stressed the importance of self-destruction in case of a miss.

The memo also mentions that Einstein, in his letter of July 5 to Bardeen and Brown, proposed another arrangement: a primary and a secondary coil perpendicular to each other. This gives zero mutual inductance without critical adjustments. The problem with flat-bottom ships remains, but if the signal under the edge of the ship were sufficiently large, "the absence of signal under the middle actually might be an advantage." They proposed testing these suggestions in model measurements, and to keep Einstein informed of the induction exploder program of the laboratory.

In his letter of August 12, Brunauer announced that the Naval Ordnance Laboratory was testing Einstein's two suggestions regarding torpedoes and submitted a report.[120] It is not available.

In mid-July, Brunauer announced that he would visit Einstein with John von Neumann on July 19,[121] and Einstein was prepared to receive them.[122]

Neumann's interest in hydrodynamic turbulence and theory of shocks was fueled by problems of nonlinear partial differential equations. Before joining the Los Alamos Scientific Laboratory to work on the atomic bomb, he was member of the navy's Bureau of Ordnance.[123] Neumann's publication list for 1941–44 mentions reports to the Bureau of Ordnance on shock waves; apparently he was asked to discuss shock waves with Einstein in general terms, but Brunauer's next letter (of July 28), which probably discussed details of the meeting, is not extant; it is mentioned only in Einstein's answer of July 30.[124] From it we may infer that Brunauer must have requested his opinion on the action of a torpedo on the armored hull of a ship. Einstein announced that he was able to simplify the mathematical description of the underwater explosion. He had already discussed this idea with John von Neumann and was happy to cooperate with him. This discussion happened in Brunauer's absence; either Einstein met Neumann on July 16 without Brunauer being present, or, if Brunauer was present on July 16, Einstein mentioned a second meeting that took place between July 16 and 30. At one of their meetings, Einstein expounded to Neumann his idea on boostering torpedo warheads.

Brunauer was eager to learn details,[125] and Einstein gave them in his letter of August 22.[126] The gist of the idea is that the best result can be achieved if the torpedo explodes when its head and not its rear reaches the ship's hull.

The pressure exerted by an explosion that takes place in E (*fig.* 2 in fig. 5.15) will give velocities to the parts of an elastic wall, which, by its plastic deformation, will absorb their kinetic energy. If the deformation is deep enough, the plate will break.

The size of the deformed part grows with h, and, as a consequence of the dispersion of the same energy over a larger area, the probability of producing a hole will diminish. Upon Einstein's request, pictures of such damages were presented by Lieutenant Commander Roy W. Goranson, who visited Einstein on August 9.[127] These pictures often showed a central hole on the sideplate of

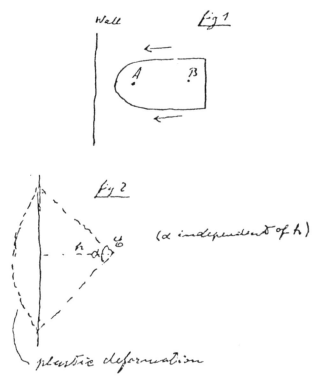

Figure 5.15. Torpedo blast.
Einstein to Stephen Brunauer, August 22, 1943. Courtesy Albert
Einstein Archives, The Hebrew University of Jerusalem.

a ship with radial ruptures, indicating that the maximum deformation took
place at the center.

Einstein proposed two explosions: a weaker one to make a central hole, then
another one to produce radial ruptures. If the first explosion were launched at
the rear of the torpedo, it would represent a larger h, and the energy would not
be enough to make the central hole.

He also proposed a hollow tip for the torpedo "to ensure a perforation" (fig.
5.16). Schwarz guesses that the tip should possibly be armed with a projectile,
to increase the likelihood of puncturing the hull instead of just shaking it.[128]
In a second figure Einstein sketched the same hollow tip with a disk in it that
can move freely and can hit the head of the tip upon impact.

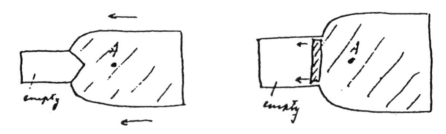

Figure 5.16. Torpedo head.

Einstein to Stephen Brunauer, August 22, 1943. Courtesy Albert Einstein Archives, The Hebrew University of Jerusalem.

Brunauer was delighted to read Einstein's new suggestion, and proposed to put it in mathematical form by making use of Kirkwood's theory of damage to diaphragms. At the same time, he announced that he would like to visit Einstein on September 4 together with Neumann.[129]

Einstein, however, was of the opinion that more calculations would not help solve the problem, for to simplify the task one has to introduce doubtful assumptions, which may substantially influence the outcome. "Experiment seems to me the only reliable way of confirmation," he stressed, and praised Kirkwood as the most experienced specialist for the enterprise.[130]

On August 13, Einstein mentioned to Brunauer a certain method of protection against torpedoes that he had suggested to Goranson: the opposite of the problem of how to make the greatest damage by torpedo.[131] We do not have details of this discussion, only Goranson's follow-up letter on research that had already been made in energy-absorbing systems. Unfortunately the figure that would make his explanation understandable is missing.

Einstein had been requested by Vannevar Bush to accept an appointment as consultant to Division 8 of the National Defense Research Committee. He asked Brunauer's opinion of whether this offer was compatible with his navy appointment. Brunauer confessed that he himself had suggested the invitation to Bush.[132]

In October 1943 Einstein proposed a trick for boosting a torpedo when it is in parallel contact with the ship, for he considered it would cause more damage than if boosted from the front or rear.[133] The velocity of the water increases as the torpedo approaches the ship, for it is dragged by the ship, so the

Figure 5.17. Torpedo head.
Einstein to Stephen Brunauer, January 4, 1944. Courtesy
Albert Einstein Archives, The Hebrew University of
Jerusalem.

torpedo head will turn in the direction of the ship. The propeller of the torpedo will keep its head contacting the wall until its position turns parallel to it; then it starts moving away from it. This is the time the torpedo should be boosted. The explosion can be initiated by a device that acts when a certain point of the torpedo contacts the ship's hull; another option is to use a time fuse that ignites when the head contacts it. A third option is the combination of these two: a contact device that ignites a time fuse when the torpedo has already made an acute angle with the hull.

The problem was eventually solved by army engineers by redesigning the contact detonator's firing pin.[134]

Even though Einstein had a slight flu,[135] he met with Brunauer (now promoted to lieutenant commander) on January 3, 1944, and brought up the idea to stop the torpedo parallel to the wall of the ship before explosion. The next day, however, he realized that the idea would not work, because the force needed to stop a torpedo of 100 kilograms (220 pounds) having a velocity of 25 meters per second (56 miles per hour) would be 300 tons (660,000 pounds) which would certainly destroy it. The torpedo must therefore explode before loosing its speed, before its essential parts undergo deformation. If the head had a hollow front with a depth of 10 centimeters (about 4 inches) (fig. 5.17), its crumpling would add 0.004 seconds to the time elapsing between its impact with the hull and the explosion.[136]

Einstein's next extant letter is addressed to Brunauer and George Gamow.[137] Apparently as a part of a longer discussion, in this letter of a mere six lines and a drawing, Einstein proposed an initiator that, by its inertia, is almost equivalent to a rigid wall. "The location of initiation and the gas capsule is chosen in such a way that the compression of the latter is effected in the shortest time." With it, he hoped to ensure a quick compression (fig. 5.18).

By indicating "lead or steel" as the arena of the explosion, apparently this time he did not discuss torpedoes but perhaps mines, underwater bombs used to destroy shore facilities and ships. This is confirmed by Gamow's reminiscences. He was picked to visit Einstein biweekly with a thick suitcase

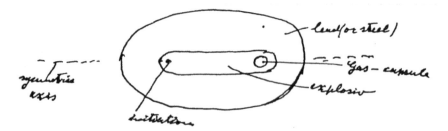

Figure 5.18. Initiator for sea mines?

Einstein to Stephen Brunauer and George Gamow, October 15, 1944. Courtesy Albert Einstein Archives, The Hebrew University of Jerusalem.

full of confidential documents. "There was a great variety of proposals, such as exploding a series of underwater mines placed along a parabolic path that would lead to the entrance of a Japanese Naval Base, with 'follow up'—aerial bombs to be dropped on the flight decks of Japanese Aircraft carriers. Einstein would meet me in his study at home, . . . and we would go through all the proposals, one by one. He approved practically all of them, saying ' . . . Oh yes, very interesting, very, very, ingenious'—and the next day the admiral in charge of the bureau was very happy when I reported to him Einstein's comments."[138]

Brunauer's memories are different: "Gamow, in later years, gave the impression that he was the Navy's liaison man with Einstein, that he visited every two weeks, and that the professor 'listened' but made no contribution— all false. The greatest frequency of visits was mine, and that was about every two months."[139]

Einstein did not spend much time on his war project. When Brunauer asked him,[140] he gave five days for three months.[141] Brunauer considered it too low, and offered a day per week, that is, thirteen days from June 1 to August 31.[142] Einstein accepted the offer.[143] Einstein's contract with the navy is unavailable. If it was for a year (Einstein signed another contract with the Bureau of Ordnance for a year from July 1, 1945 to June 30, 1946, but without any obligation for consultation),[144] then he earned for these two years around $2,500, an annual sum of around $1,250—one-tenth of his salary ten years earlier, when he joined the Institute for Advanced Study.

"It is not appropriate to cite specific developments as having been his," Brunauer noted, evaluating Einstein's contributions. "The new and more effective high explosives developed during the war resulted from teamwork, and Einstein was a member of the team. He also made another contribution—to morale. It was uplifting to know that 'Einstein was one of us.' " [145]

This was true for all his contributions to technology. He always worked as part of a team whose members, whether few or many, felt Einstein stood beside them.

Notes

Preface

1. В. Я. Френкель and Б. Е. Явелов, *Эйнштейн: изобретения и эксперимент*, 2. изд. (Москва: Наука, 1990). A short account was published as Viktor Frenkel and Boris Yavelov, "'What May Happen to a Man Who Thinks a Great Deal but Reads Very Little,'" in Yuri Balashov and Vladimir Vizgin, eds., *Einstein Studies in Russia* (Boston: Birkhäuser, 2002), pp. 297–306.

2. Karl Wolfgang Graff, "Albert Einstein als Erfinder in den Jahren 1907–1933" (Thesis, Historisches Institut der Universität Stuttgart, 2004).

3. Michael Eckert, *The Dawn of Fluid Dynamics: A Discipline between Science and Technology* (Weinheim, Germany: Wiley-VCH, 2006), pp. 57–81.

4. Jobst Broelmann, *Intuition und Wissenschaft in der Kreiseltechnik. 1750 bis 1930* (Munich: Deutsches Museum, 2002).

5. Dieter Lohmeier and Bernhardt Schell, eds., *Einstein, Anschütz und der Kieler Kreiselkompass—Einstein, Anschütz and the Kiel Gyro Compass*, 2nd ed. (Kiel, Germany: Raytheon Marine GmbH, 2005).

6. Matthew Trainer, "Albert Einstein's Patents," *World Patent Information* 28 (2006): 159–65.

7. I. Rosenberg to Einstein, May 24 and June 19, 1919, *Collected Papers of Albert Einstein* (*CPAE*; Princeton, NJ: Princeton University Press, 1987–2009), vol. 9, pp. 567 and 570, resp.

8. Einstein to I. Rosenberg, May 25, 1919, *CPAE*, vol. 9, p. 567.

9. Hermann Anschütz-Kaempfe to Einstein, Aug. 20, 1922, Albert Einstein Archives, Jerusalem (AEA), [37 382].

10. Einstein to Elsa Einstein, May 4, 1922, AEA, [143 124].

11. Einstein to Elsa Einstein, Aug. 14, 1923, AEA, [143 127].

12. Lohmeier and Schell, *Einstein, Anschütz und der Kieler Kreiselkompaß*.

13. [Max Born, ed.,] *The Born-Einstein Letters, 1916–1955: Friendship, Politics and Physics in Uncertain Times* (Houndmills: Macmillan, 2005), p. 138.

Introduction

1. For details on the Einstein company, see Nicolaus Hettler, "Die Elektrotechnische Firma J. Einstein u. Cie in München–1876–1894" (Dissertation, Universität Stuttgart, 1996). Stefan Siemer's "'In the Brightest Arc Lamps and Incandescent Lights': The Electrical Factory *Jakob Einstein und Cie*" is a short but well-illustrated history of the company, in Jürgen Renn, ed., *Albert Einstein Chief Engineer of the*

Universe: One Hundred Authors for Einstein (Weinheim, Germany: Wiley-VCH Verlag, 2005), pp. 128–33.

2. Otto Neustätter to Einstein, Mar. 12, 1929, Albert Einstein Archives, Jerusalem (AEA), [30 436].

3. Einstein to Mileva Marić, Sept. 6, 1900, *Collected Papers of Albert Einstein* (*CPAE;* Princeton, NJ: Princeton University Press, 1987–2009), vol. 1, doc. 74.

4. "Autobiographisches–Autobiographical Notes" (1946), in Paul A. Schilpp, ed., *Albert Einstein Philosopher-Scientist* (Evanston, IL: Library of Living Philosophers, 1949), pp. 14–15; "Autobiographische Skizze" (1955), in Carl Seelig, ed., *Helle Zeit-dunkle Zeit. In memoriam Albert Einstein* (Zurich: Europa, 1956), p. 10.

5. Ibid., p. 12.

6. Friedrich Haller to Bundesrat, Sept. 5, 1904, AEA, [70 075].

7. Max Flückiger, *Albert Einstein in Bern: das Ringen um ein neues Weltbild; eine dokumentarische Darstellung über den Aufstieg eines Genies* (Bern: Haupt, 1974), p. 65. English translation from *CPAE*, vol. 5, doc. 34, n. 1.

8. Einstein to Heinrich Zangger, before Aug. 11, 1918, *CPAE*, vol. 8, doc. 597. He wrote to Mileva Einstein-Marić in a similar vein around Nov. 10, 1918, *CPAE*, vol. 8, doc. 648.

9. A. Einstein and Jakob Grommer, "Beweis der Nichtexistenz eines überall regulären zentrisch-symmetrischen Feldes nach der Feld-Theorie von Th. Kaluza," *Scripta Universitatis Bibliothecae Hierosolymitanarum. Mathematica et Physica* 1 (1923): no. 7.

10. Einstein to Franz Rusch, Mar. 18, 1921, *CPAE*, vol. 12, doc. 105.

11. Einstein to Heinrich Zangger, before Aug. 11, 1918, *CPAE*, vol. 8, doc. 597.

12. Einstein to Arnold Sommerfeld, Feb. 5, 1919, *CPAE*, vol. 9, doc. 5.

13. Einstein to Hermann Weyl, June 6, 1922. AEA, [24 071].

14. Einstein to Michele Besso, Dec. 12, 1919, *CPAE*, vol. 9, doc. 207.

15. Einstein to Friedrich Kottler, July 29, 1920, *CPAE*, vol. 10, doc. 88.

16. Spencer R. Weart, and Gertrud Weiss Szilárd, eds., *Leo Szilard: His Version of the Facts: Selected Recollections and Correspondence* (Cambridge, MA: MIT Press, 1978), pp. 9–12.

17. Einstein to Heinrich Zangger, before Aug. 11, 1918, *CPAE*, vol. 8, doc. 597.

18. Einstein to Otto Marx, Dec. 22, 1917, *CPAE*, vol. 6, doc. 415.

19. Einstein to Chester A. Clark, after June 17, 1921, *CPAE*, vol. 12, doc. 150.

20. Einstein to Michele Besso, before July 26, 1920, *CPAE*, vol. 10, doc. 85.

21. Gerald Holton, *Thematic Origins of Scientific Thought: Kepler to Einstein* rev. ed. (Cambridge, MA: Harvard University Press, 1988), p. 395.

22. Thomas P. Hughes, "Einstein, Inventors, and Invention," *Science in Context* 6 (1993): 25–42.

CHAPTER 1: Musings

1. G. B. Seybold, "A Sailing Ship without Sails: New Wonder of the Seas," *Popular Science Monthly*, Feb. 1925, www.rexresearch.com/flettner/flettner.htm.

2. Albert Einstein Archives, Jerusalem (AEA), [29 132].

3. Albert Einstein, "El buque de Flettner," *La Prensa* (Buenos Aires), Apr. 13, 1925, p. 10. Its original German manuscript is in AEA, [1 049].

4. Ludwig Prandtl, "Magnuseffekt und Windkraftschiff," *Die Naturwissenschaften* 13 (1925): 93–108.

5. Jakob Ackeret, *Das Rotorschiff und seine physikalischen Grundlagen* (Göttingen: Vanderholdt & Ruprecht, 1925).

6. For Flettner's way to the rotor ship and his cooperation with Prandtl's institute, see Anton Flettner, *Mein Weg zum Rotor* (Leipzig, Germany: Koehler & Amelang, 1926).

7. "America's First Rotor Boat," *Popular Science Monthly*, Sept. 1925, www.rex research.com/flettner/flettner.htm.

8. C. P. Gilmore, "Spin Sail Harnesses Mysterious Magnus Effect for Ship Propulsion," *Popular Science*, Jan. 1984, www.rexresearch.com/flettner/flettner.htm.

9. DE 10 2006 025 732 A1 2007.12.06: Rolf Rohden, "Magnusrotor," 2007.

10. Albert Einstein, "Die Ursache der Mäanderbildung der Flußläufe und des sogenannten Baerschen Gesetzes," *Die Naturwissenschaften* 14 (1926): 223–24. English translation from Albert Einstein, *Ideas and Opinions* (New York: Crown, 1982), p. 249.

11. Harold Mottsmith and Irving Langmuir, "Radial Flow in Rotating Liquids," *Physical Review* 20 (1922): 95.

12. James Thomson, "On the Origin of Windings of Rivers in Alluvial Plains, with Remarks on the Flow of Water round Bends in Pipes," *Proceedings of the Royal Society of London* 25 (1876–77): 5–8.

13. Ludwig Prandtl, "Bemerkung zu dem Aufsatz von A. Einstein 'Die Ursache . . . ,'" *Die Naturwissenschaften* 14 (1926): 619–20.

14. J. Isaachsen, "Über einige Wirkungen von Zentrifugalkräften in Flüssigkeiten und Gasen," *Civilingenieur* 42 (1896): 353–86; "Innere Vorgänge in strömenden Flüssigkeiten und Gasen," *Zeitschrift des Vereines der deutschen Ingenieure* 55 (1911): 215–21.

15. [Max Born, ed.,] *The Born-Einstein Letters, 1916–1955: Friendship, Politics and Physics in Uncertain Times* (Houndmills: Macmillan, 2005), p. 138.

16. Ibid., p. 140.

17. D. Montgomery, "On Stirring a Cup of Tea," *American Journal of Physics* 19 (1951): 477; J. Satterly, "Rotating Liquid Motions," *American Journal of Physics* 27 (1959): 526.

18. R. A. Alpher and R. Herman, "Tea Leaves, Baer's Law, and Albert Einstein," *American Journal of Physics* 28 (1960): 5–8.

19. Harold Mottsmith and Irving Langmuir, "Radial Flow in Rotating Liquids," *Physical Review* 20 (1922): 95.

20. Kent A. Bowker, "Albert Einstein and Meandering Rivers," http://searchand discovery.net/documents/Einstein/albert.htm, accessed May 16, 2008. Originally published in *Earth Science History* 7 (1988): 45.

21. S. A. Schumm, ed., *River Morphology* (Stroudsburg, PA: Dowden, Hutchinson and Ross, 1972), p. 234.

22. Andrew S. Goudie, "Baer's Law of Stream Deflection," *Earth Sciences History* 23 (2004): 278–82.

23. A. E. Scheidegger, *Theoretical Geomorphology* (Berlin: Springer, 1961).

24. Jesus Martinez-Frias, David Hochberg, and Fernando Rull, "Contributions of Albert Einstein to Earth Sciences: A Review in Commemoration of the World Year of Physics," http://arxiv.org/fkp/physics/papers/0512114.pdf, accessed May 17, 2008.

25. T. B. Liverpool and S. F. Edwards, "The Dynamics of a Meandering River," *Physical Review Letters* 75 (1995): 3016–19; F. Noisy, T. Pastutto, G. Gauthier, P. Gondret, and M. Rabaud, "Instabilités spirales entre disques tournants," *Bulletin S. F. P.* 135 (2001): 4–8.

26. Stephen Pincock, "Einstein's Tea-Leaves Inspire New Gadget," *ABC Science Online*, Jan. 17, 2007, www.abc.net.au/science.

27. Dian R. Arifin, Leslie Y. Yeo, and James R. Friend, "Microfluidic Blood Plasma Separation via Bulk Electrodynamic Flows," *Biomicrofluidics* 1, 014103 (2007): 1–13.

28. Andrew S. Goudie, "Baer's Law of Stream Deflection," *Earth Sciences History* 23 (2004): 278–82.

CHAPTER 2: Experiments

1. Albert Einstein, "Theoretische Bemerkungen zur Supraleitung der Metalle," in *Het natuurkundig laboratorium der Rijksuniversiteit te Leiden in de jaren 1904–1922. Gedenkboek aangeboden aan H. Kamerling Onnes, Directeur van het laboratorium, bij gelegenheid van zijn veertigjarig professoraat op 11 november 1922* (Leiden, the Netherlands: Ijdo, 1922), p. 429.

2. *Kaizo* 5, no. 2 (Feb. 1923): 1–7.

3. Einstein to Mileva Marić, Sept. 10 and 28(?), 1899, *Collected Papers of Albert Einstein* (*CPAE*; Princeton, NJ: Princeton University Press, 1987–2009), vol. 1, docs. 54 and 57, resp.

4. Einstein to Wilhelm Wien, July 10, 1912, *CPAE*, vol. 5, doc. 413.

5. Roland Eötvös, "Über die Anziehung der Erde auf verschiedene Substanzen," *Mathematische und naturwissenschaftliche Berichte aus Ungarn* 8 (1891): 65–68.

6. Albert Einstein, "Zur Elektrodynamik bewegter Körper," *Annalen der Physik* 17 (1905): 891–921, *CPAE*, vol. 2, doc. 23.

7. Walter Kaufmann, "Die magnetische und elektrische Ablenkbarkeit der Becquerelstrahlen und die scheinbare Masse der Elektronen," *Königliche Gesellschaft der Wissenschaften zu Göttingen. Mathematisch-physikalische Klasse. Nachrichten* (1901): 143–55; "Über die Konstitution des Elektrons," *Königlich Preußische Akademie der Wissenschaften* (Berlin), *Sitzungsberichte* (1905): 949–56.

8. Albert Einstein, "Über eine Methode zur Bestimmung des Verhältnisses der transversalen und longitudinalen Masse des Elektrons," *Annalen der Physik* 21 (1906): 3–56, *CPAE*, vol. 2, doc. 36.

9. Albert Einstein, and Wander J. de Haas, "Experimenteller Nachweis der Ampèreschen Molekularströme," *Deutsche Physikalische Gesellschaft, Verhandlungen* 17 (1915): 152–70, *CPAE*, vol. 6, doc. 13. Published in English as Albert Einstein and Wander J. de Haas, "Experimental Proof of the Existence of Ampère's Molecular Currents," *Koninklijke Akademie van Wetenschappen te Amsterdam. Section of Sciences. Proceedings* 1 (1915–16): 696–711, *CPAE*, vol. 6, doc. 14. These papers are summarized in Albert Einstein, "Experimenteller Nachweis der Ampèreschen Molekularströme," *Die Naturwissenschaften* 3 (1915): 237–38, *CPAE*, vol. 6, doc. 15.

10. Albert Einstein, "Ein einfaches Experiment zum Nachweis der Ampèreschen Molekularströme," *Deutsche Physikalische Gesellschaft. Verhandlungen* 18 (1916): 173–77, *CPAE*, vol. 6, doc. 28.

11. Peter Galison, *How Experiments End* (Chicago: University of Chicago Press, 1986), pp. 47–52.

12. Einstein to Emile Meyerson, Jan. 27, 1930, Albert Einstein Archives, Jerusalem (AEA), [35 384].

13. Einstein and Wander J. de Haas, "Experimenteller Nachweis der Ampèreschen Molekularströme," *Deutsche Physikalische Gesellschaft, Verhandlungen* 17 (1915): 152–70, esp. 153, 156, *CPAE*, vol. 6, doc. 13.

14. Einstein to Wander de Haas, [1922–28], AEA, [70 414].

15. J. N. Fisher, "Some Further Experiments on the Gyromagnetic Effect," *Proceedings of the Royal Society* A 109 (1925): 7–27.

16. *Physikalische Berichte* 8 (1922): 1400–1401.

17. Peter Pringsheim to Einstein, Aug. 17, 1924, AEA, [19 146].

18. Albert Einstein, "Schallausbreitung in teilweise dissoziierten Gasen," *Preußische Akademie der Wissenschaften* (Berlin), *Sitzungsberichte* (1920): 380–85, *CPAE*, vol. 7, doc. 39.

19. Friedrich Keutel, "Ueber die spezifische Wärme der Gasen" (Thesis. Berlin: Ebering, 1910).

20. Eduard Grüneisen, and Erich Goens, "Schallgeschwindigkeit in Stickstofftetroxyd. Eine untere Grenze seiner Dissoziationsgeschwindigkeit," *Annalen der Physik* 72 (1923): 193–220.

21. Arend J. Rutgers, "Zur Dispersionstheorie des Schalles," *Annalen der Physik* 16 (1933): 350–59; Karl F. Herzfeld, "Fifty Years of Physical Ultrasonics," *Journal of the Acoustical Society of America* 39 (1966): 1–11.

22. Hermann Anschütz-Kaempfe to Einstein, Dec. 28, 1920, *CPAE*, vol. 10, doc. 247; Einstein to Hermann Anschütz-Kaempfe, Oct. 11, 1921, *CPAE*, vol. 12, doc. 263; Max Schuler to Einstein, Nov. 7, 1921, *CPAE*, vol. 12, doc. 290; Einstein to Max Schuler, Dec. 1, 1921, *CPAE*, vol. 12, doc. 309.

23. Dieter Lohmeier and Bernhardt Schell, eds., *Einstein, Anschütz und der Kieler Kreiselkompass–Einstein, Anschütz and the Kiel Gyro Compass*, 2d ed. (Kiel: Raytheon Marine GmbH, 2005), p. 39.

24. Einstein to Max Schuler, Dec. 1, 1921, *CPAE*, vol. 12, doc. 309.

25. Einstein to Hermann Anschütz-Kaempfe, Jan. 9, 1922, AEA, [81 205].

26. Einstein to Hermann Anschütz-Kaempfe, Sept. 17, 1921, *CPAE*, vol. 12, doc. 237.

27. Einstein to Hermann Anschütz-Kaempfe, June 18, 1922, AEA, [80 280].

28. Einstein to Hendrik A. Lorentz, Jan. 23, 1915, *CPAE*, vol. 8, doc. 47.

29. Hans Mühsam to Einstein, Aug. 4, 1942, AEA, [38 339].

30. AEA, [17 078].

31. Hermann Mark to Peter Bergmann, Apr. 11, 1967, AEA, [17 080].

32. Albert Einstein, "Zur affinen Feldtheorie." *Preußische Akademie der Wissenschaften* (Berlin), *Sitzungsberichte* (1923): 137–40.

33. Hermann Mark to Peter Bergmann, Apr. 11, 1967, AEA, [17 080].

34. W. F. G. Swann and A. Longacre, "An Attempt to Detect a Magnetic Field Resulting from the Rapid Rotation of a Copper Sphere," *Physical Review* 31 (1928): 1115–16.

35. Einstein to Hendrik A. Lorentz, Jan. 1, 1921, *CPAE*, vol. 12, doc. 3.

36. Einstein to Paul Ehrenfest, Jan. 20, 1921, *CPAE*, vol. 12, doc. 24.

37. Einstein to Max Born, Jan. 31, 1921, *CPAE*, vol. 12, doc. 37.

38. Max Born to Einstein, Feb. 12, 1921, *CPAE*, vol. 12, doc. 47.

39. Paul Ehrenfest to Einstein, Jan. 22, 1921, *CPAE*, vol. 12, doc. 30.

40. Einstein to Max Born, Aug. 22, 1921, *CPAE*, vol. 12, doc. 211.

41. Einstein to Arnold Sommerfeld, Sept. 27, 1921, *CPAE*, vol. 12, doc. 247.

42. Einstein to Heinrich Zangger, Sept. 29, 1921, *CPAE*, vol. 12, doc. 249; Einstein to Michele Besso, Oct. 20, 1921, *CPAE*, vol. 12, doc. 275.

43. Albert Einstein, "Über ein den Elementarprozeß der Lichtemission betreffendes Experiment," *Preußische Akademie der Wissenschaften* (Berlin), *Sitzungsberichte* (1921): 882–83, *CPAE*, vol. 7, doc. 68. Submitted Dec. 6, 1921.

44. For a reconstruction of what he might have in mind on the Doppler shift of canal rays, see note 5 to Einstein's letter to Arnold Sommerfeld, Sept. 9, 1921, *CPAE*, vol. 12, doc. 261.

45. Hans Geiger to Einstein, Nov. 7, 1921, *CPAE*, vol. 12, doc. 289.

46. Walther Bothe to Einstein, Dec. 7, 1921, *CPAE*, vol. 12, doc. 316.

47. Albert Einstein, Hans Geiger, Walther Bothe, "Über ein optisches Experiment, dessen Ergebnis mit der Undulationstheorie unvereinbar ist," incomplete manuscript of a lecture delivered to the Prussian Academy on Jan. 19, 1922, AEA, [2 086], announced in *Preußische Akademie der Wissenschaften* (Berlin), *Sitzungsberichte* (1922): 2.

48. Einstein to Hedwig and Max Born, Dec. 30, 1921, *CPAE*, vol. 12, doc. 345.

49. Arnold Sommerfeld to Einstein, Jan. 11, 1922, AEA, [21 347].

50. Paul Ehrenfest to Einstein, Jan. 19 and 26, 1922, AEA, [10 009] and [10 013].

51. As Einstein mentioned in his letters to Arnold Sommerfeld on or after Jan. 18, 1922, AEA, [21 402], and Paul Ehrenfest, between Jan. 19 and 22, 1922, AEA, [10 011].

52. Einstein to Paul Ehrenfest, [Jan. 26, 1922], AEA, [10 015].

53. Albert Einstein, "Zur Theorie der Lichtfortpflanzung in dispergierenden Medien," *Preußische Akademie der Wissenschaften* (Berlin), *Sitzungsberichte* (1922): 18–22.

54. Einstein to Hans Albert and Eduard Einstein, Feb. 12, 1922, AEA, [75 615].

55. Einstein to Paul Ehrenfest, Feb. 12, 1922, AEA, [10 019].

56. Einstein to Paul Ehrenfest, between Jan. 19 and 22, 1922, AEA, [10 011].

57. W. Orthmann to Peter Pringsheim, and Pringsheim to Einstein, Nov. 12, 1923, AEA, [19 140].

58. Peter Pringsheim to Einstein, July 19, 1922, AEA, [19 138].

59. Peter Pringsheim to Einstein, Nov. 17, 1923, AEA, [19 144].

60. Arnold Sommerfeld to Arthur Compton, Oct. 9, 1923, in Michael Eckert and Karl Märker, eds., *Arnold Sommerfeld Wissenschaftliche Briefwechsel Band 2: 1919–1951* (Berlin: Deutsches Museum Verlag für Geschichte der Naturwissenschaften und der Technik, 2004), p. 153. He also mentioned to Charles E. Mendenhall that when he spent a few days with Einstein, they occupied most of the time discussing Compton's experiment (see his letter to Niels Bohr, Jan. 21, 1923, ibid., p. 144).

61. Arthur Compton, "A Quantum Theory of the Scattering of X-Rays by Light Elements," *Physical Review* 21 (1923): 207.

62. Arthur Compton, and Samuel Allison, *X-Rays in Theory and Experiment*, 2nd ed. (New York: Van Nostrand, 1935), p. 48.

63. "Das Komptonsche Experiment. Ist die Wissenschaft um ihrer selbst wissen da?" *Berliner Tageblatt*, Apr. 20, 1924, Morgen Express Ausgabe, 1. Beiblatt.

64. Hermann Mark to Einstein, Sept. 14, 1923, AEA, [17 076].

65. Peter Debye, "Zerstreuung von Röntgenstrahlen und Quantentheorie," *Physikalische Zeitschrift* 24 (1923): 161–66.

66. G. E. M. Jauncey, and Carl H. Eckart, "Is There a Change of Wave-Length on Reflection of X-rays from Crystals?" *Nature* 112 (1923): 325–26.

67. Hermann Mark to Einstein, Sept. 28, 1923, AEA, [17 077].

68. Arthur Compton,"The Spectrum of Scattered X-Rays," *Physical Review* 22 (1923): 409–13.

69. Hermann Mark to Peter G. Bergmann, Apr. 11, 1967, and to Gerald Holton, Jan. 9, 1986, AEA, [17 080] and [73 352], resp.

70. Hermann Mark, "Der Comptoneffekt. Seine Entdeckung und seine Deutung durch die Quantentheorie," *Die Naturwissenschaften* 30 (1925): 494–500; H. Kallmann and Hermann Mark, "Der Comptonsche Streuprozeß," *Ergebnisse der exakten Naturwissenschaften* 5 (1926): 267–325.

71. Albert Einstein, "Vorschlag zu einem die Natur des elementaren Strahlungs-Emissionsprozesses betreffenden Experiment," *Die Naturwissenschaften* 14 (1926): 300–301.

72. Emil Rupp, "Interferenzuntersuchungen an Kanalstrahlen," *Annalen der Physik* 79 (1926): 1–34.

73. Einstein to Emil Rupp, Mar. 20, 1926, AEA, [70 701].

74. Paul Ehrenfest to Einstein, Apr. 7, 1926, AEA, [10 134].

75. Albert Einstein, "Über die Interferenzeigenschaften des durch Kanalstrahlen emittierten Lichtes," *Preußische Akademie der Wissenschaften* (Berlin), *Sitzungsberichte* (1926): 334–40.

76. Emil Rupp, "Über die Interferenzeigenschaften des Kanalstrahllichts," *Preußische Akademie der Wissenschaften* (Berlin), *Sitzungsberichte* (1926): 341–51.

77. For the history and a detailed analysis of Einstein's collaboration with Rupp, see Jeroen van Dongen, "Emil Rupp, Albert Einstein and the Canal Ray Experiments on Wave-Particle Duality: Scientific Fraud and Theoretical Bias," *Historical Studies in the Physical and Biological Sciences* 37 suppl. (2007): 73–120.

78. Jeroen van Dongen, "The Interpretation of the Einstein-Rupp Experiments and Their Influence on the History of Quantum Mechanics," *Historical Studies in the Physical and Biological Sciences* 37 suppl. (2007): 121–30.

79. Einstein's theoretical and experimental considerations in the field of super-conductivity were reconstructed from scattered sources by Tilman Sauer in his comprehensive paper "Einstein and the Early Theory of Superconductivity, 1919–1922," *Archive for History of Exact Sciences* 61 (2007): 159–211. This chapter relies on it.

80. Einstein to Paul Ehrenfest, Sept. 2, 1921, *CPAE*, vol. 12, doc. 225.

81. Einstein to Hendrik A. Lorentz, Jan. 1, 1921, *CPAE*, vol. 12, doc. 3.

CHAPTER 3: Expert Opinions

1. *Collected Papers of Albert Einstein* (*CPAE*; Princeton, NJ: Princeton University Press,1987–2009), vol. 5, doc. 67. The main patent is CH38853: "Wechselstromkollektor-

maschine mit Kurzschlußbürsten und diesen gegenüberliegenden Hilfspulen zur Funkenvermeidung." The patent application examined by Einstein was eventually published as CH39988: "Wechselstromkollektormaschine mit Kurzschlußbürsten und diesen gegenüberliegenden Hilfspulen zur Funkenvermeidung. Zusatzpatent zum Hauptpatent 38853."

2. Schweizerisches Bundesarchiv, Bern, Switzerland, E 22/2338, Dossier Einstein, AEA, [72 270].

3. CH35840: Ignacy Mościcki, "Apparat zur Erzeugung von Stickstoffoxyden auf elektrischem Wege," submitted Jan. 26, 1906.

4. K. Drewnowski et al., *Profesor Dr. Ignacy Mościcki. Żicie i działaność na polu nauki i techniki* (Warsaw: Nakład komitetu uczczenia 30-lecia pracy naukowej Profesora Dr. Ignacy Mościckiego, 1934), p. 12.

5. Zofia Gołąb-Meyer, "Prezydent Rzeczypospolitej Polskiej Ignacy Mościcki i Albert Einstein," *Foton* 91 (Winter 2005): 45–47. In English as "Albert Einstein and Mościcki's Patent Application," *Physics Teacher* 44 (2006): 212–13. I owe this reference to Trevor Lipscombe.

6. Einstein to Ignacy Mościcki, Sept. 8, 1932, AEA, [20 170].

7. Ignacy Mościcki to Einstein, Sept. 27, 1932, AEA, [71 782].

8. Einstein to Ignacy Mościcki, Oct. 19, 1932, AEA, [20 176].

9. Albert Gockel's diary entry of June 28, 1908 (Marianne Baumhauer, Freising, Germany, cited in *CPAE*, vol. 5, doc. 104, n. 5); see also Einstein to Albert Gockel, Dec. 3, 1908 and Mar. 25(?), 1909, *CPAE*, vol. 5, docs. 130 and 144, resp.

10. Joseph Kowalski to Einstein, Mar. 30, 1908, *CPAE*, vol. 5, doc. 94.

11. www.ige.ch/e/institut/i1094.shtm (Swiss Federal Institute of Intellectual Property).

12. CH39853: Joseph Lemblé, "Elektrische Typenschiffchen-Schreibmaschine," 1907.

13. CH39561: Xaver Koller, "Kiessortiermaschine," 1907.

14. CH39619: Kammerer & Schneider, "Wetteranzeiger, der durch die Feuchtigkeit der Luft beeinflußt wird," 1907.

15. For a comprehensive history of Einstein's expertise in Hermann Anschütz-Kaempfe's legal cases, see the introduction to Dieter Lohmeier, and Bernhardt Schell, eds., *Einstein, Anschütz und der Kieler Kreiselkompass–Einstein, Anschütz and the Kiel Gyro Compass* 2nd ed. (Kiel, Germany: Raytheon Marine GmbH, 2005) by Bernhardt Schell.

16. DE182855: Hermann Anschütz-Kaempfe, and Friedrich von Schirach, "Kreiselapparat," 1907.

17. DE34513: Marinus Gerardus van den Bos, and Barend Janse, "Neuerung an Schiffscompassen," 1886.

18. DE236200: Anschütz & Co., "Kreiselkompaß, dessen Rotationsachsen in der Horizontalebene teilweise gefesselt ist und daher Schwingungen um die Nord-Südrichtung ausführt," 1911.

19. Felix Klein, and Arnold Sommerfeld, *Über die Theorie des Kreisels* (Leipzig, Germany: Teubner, 1897–1910).

20. Arnold Seligsohn, *Patentgesetz und Gesetz, betreffend den Schutz von Gebrauchsmustern*, 6th ed. (Berlin: De Gruyter, 1920), p. 399.

21. "Expert Opinion on Legal Dispute between Anschütz & Co. and Sperry Gyroscope Company," Feb. 6, 1915, *CPAE*, vol. 6, doc. 12.

22. "Supplementary Expert Opinion," Aug. 7, 1915, *CPAE*, vol. 6, doc. 19.

23. Einstein to Hermann Anschütz-Kaempfe, Aug. 22, 1918, *CPAE*, vol. 8, doc. 606. For a well-documented history of the lawsuit, see Lohmeier and Schell, *Einstein, Anschütz und der Kieler Kreiselkompass*, pp. 23–28.

24. It was granted as DE307847: "Einrichtung am Kreiselkompassen zur Vermeidung von Schlingerfehlern," 1918.

25. DE241637: Anschütz & Co., "Kreiselapparat," 1911.

26. Anschütz & Co. to Einstein, June 6, 1918, *CPAE*, vol. 8, doc. 559.

27. Anschütz & Co. to Einstein, June 21, 1918, *CPAE*, vol. 8, doc. 568.

28. Anschütz & Co. to Einstein, July 12, 1918, *CPAE*, vol. 8, doc. 587.

29. Albert Einstein, "Private Expert Opinion on the Objection to Patent Application G 43359 of the Society of Nautical Instruments on the Basis of Patent 241637," July 16, 1918, *CPAE*, vol. 7, doc. 11.

30. It was granted as DE307847: "Einrichtung am Kreiselkompassen zur Vermeidung von Schlingerfehlern," 1918.

31. DE 308721: Gesellschaft für Nautische Instrumente GmbH, "Einrichtung zu Kreiselkompassen zur Vermeidung von Schlingerfehlern. Zusatz zum Patent 307847," 1918, and DE 308722: "Einrichtung an Kreiselkompassen zur Vermeidung von Schlingerfehlern. Zusatz zum Patent 307847," 1918.

32. Oscar Martienssen to Einstein, Mar. 25, 1922, Albert Einstein Archives, Jerusalem (AEA), [85 064].

33. Oscar Martienssen, "Die Theorie des Kreiselkompasses," *Zeitschrift für Instrumentenkunde* 32 (1912): 309–21.

34. Oscar Martienssen, "Ein neuer Kreiselkompaß," *Zeitschrift für Instrumentenkunde* 39 (1919): 165–80.

35. Richard Grammel, *Der Kreisel. Seine Theorie und seine Anwendungen* (Berlin: Springer, 1920).

36. Einstein to Oscar Martienssen,. Mar. 27, 1922, AEA, [44 385].

37. Einstein to Hermann Anschütz-Kaempfe, Mar. 27, 1922, AEA, [80 287].

38. Oscar Martienssen to Einstein, Mar. 28, 1922, AEA, [84 219]. It was granted as DE307847: Gesellschaft für Nautische Instrumente GmbH, "Einrichtung am Kreiselkompassen zur Vermeidung von Schlingerfehlern," 1918; DE241637: Anschütz & Co., "Kreiselapparat," 1911.

39. Einstein to Elsa Einstein, Apr. 9 [8], 1922, AEA, [143 122].

40. Einstein to Maurice Solovine, Apr. 20, 1922, AEA, [80 843].

41. Albert Einstein, "Second Supplementary Expert Opinion in the Matter of the Society of Nautical Instruments vs. Anschütz & Co.," after June 9, 1922, AEA, [84 218].

42. DE211634: Hartmann & Braun AG, "Gyroskopkompaß mit mehreren je mit verschiedenen Freiheitsgraden ausgestatteten rotierenden Massen. Zusatz zum Patent 174111," 1909.

43. Hermann Anschütz-Kaempfe to Einstein, June 25, 1922, AEA, [37 378]; Einstein to Hermann Anschütz-Kaempfe, July 1, 1922, AEA, [80 288].

44. Einstein to Hermann Anschütz-Kaempfe, July 16, 1922, AEA, [80 720].

45. Lohmeier and Schell, *Einstein, Anschütz und der Kieler Kreiselkompass*, p. 34.

46. In this case I made use of Jobst Broelmann's manuscript "Anschütz, Drexler und die Kreiselbau-Gesellschaft." Deutsches Museum: Munich, n.d.

47. DE326737: Kreiselbau GmbH, "Vorrichtung zum Berichtigen der Neigungsanzeige für Fahrzeuge, insbesondere Flugzeuge," 1917.

48. Albert Einstein, "Court Expert Opinion in the Matter of Anschütz & Co. vs. Kreiselbau-Co.," July 23, 1919, *CPAE*, vol. 7, doc. 21.

49. DE301738: Anschütz & Co., "Anzeigevorrichtung für die Drehungen eines Flugzeuges um die senkrechte Achse," 1917.

50. See DE262409: Louis Marmonier, "Vorrichtung für Stabilisierung von Luftfahrzeugen, insbesondere Flugzeugen, mittels eines Kreiselpendels," 1913; DE267061: Wilhelm Wolfromm, "Lagenanzeiger für Flugzeuge mit vor einem Skalenblatt liegenden Zeigern für Angabe der Längs- und Querneigung," 1913; DE286217: Edmund Sparmann, "Selbsttätiger Kreiselstabilisator für Flugzeuge," 1915.

51. Hugo Licht to Wolfgang Otto, Sept. 20, 1919, Archiv, Anschütz & Co. GmbH/Raytheon Marine GmbH, Kiel, Germany M 819a/I.

52. Ibid.

53. Hugo Licht to Hermann Anschütz, November 5, 1921, Archiv, Anschütz & Co. GmbH/Raytheon Marine GmbH, Kiel, Germany, M 819b.

54. Th. Rosenbaum, "Zur Theorie des Kreisels," *Schiffbau* 12 (1910–11): 115–23.

55. Hugo Licht to Hermann Anschütz, Nov. 5, 1921, Archiv, Anschütz & Co. GmbH/Raytheon Marine GmbH, Kiel, Germany, M 819b.

56. Max Schuler to Richard Grammel, Nov. 1, 1920, Archiv, Anschütz & Co. GmbH/Raytheon Marine GmbH, Kiel, Germany, M 1185.

57. Ludwig Prandtl to Max Schuler, Aug. 11, 1920, Archiv, Anschütz & Co. GmbH/Raytheon Marine GmbH, Kiel, Germany, M 1112.

58. Albert Einstein, "Response to the Expert Opinion of Hans Wolff in the Legal Dispute between Anschütz & Co and Kreiselbau GmbH," Jan. 18, 1922, AEA, [79 227].

59. GB125096: James B. Henderson, "Apparatus for Indicating Changes in the Course of a Ship, Airship, Aeroplane or the Like, also Applicable for Automatic Steering," 1919, application submitted 1916.

60. Lohmeier and Schell, *Einstein, Anschütz und der Kieler Kreiselkompass*, p. 30.

61. Albert Einstein, "Reply to the Plaintiff's Written Statement of 27. December 1916," *CPAE*, vol. 6, doc. 44.

62. Mercur Flugzeugbau GmbH to Einstein, Dec. 29, 1917, *CPAE*, vol. 8, doc. 422.

63. Alfred Zehder to Mercur Flugzeugbau, Dec. 4, 1917, AEA, [35 350].

64. DE269498: Allgemeine Elektrizitäts-Gesellschaft, "Verfahren zur Herstellung vom Wolframdrähten für Glühkörper elektrischer Glühlampen," 1914.

65. DE297015: Konrad Sannig & Co.,GmbH., "Verfahren zur Herstellung von Draht aus Wolfram und Wolframlegierungen durch Pressen des Ausgangsmaterials zu Stangen, Glühen dieser Stangen zum Zwecke des Zusammenbackens der Wolframpartikel bei 1000° bis 1200° und Zusammensintern durch Stromwärme im Wasserstoffstrome und darauffolgende mechanische Bearbeitung," 1917.

66. Albert Einstein, "Expert Opinion on German Patent 269 498 of the A.E.G., Berlin, on a 'Method for the Production of Tungsten Wires for Filaments in Incandescent Lamps,'" Jan. 10, 1920, *CPAE*, vol. 7, doc. 30.

67. Konrad Sannig & Co., GmbH, to Einstein, Nov. 18, 1920, *CPAE*, vol. 10, Calendar; AEA, [35 371].

68. Einstein to Konrad Sannig & Co., GmbH, after Nov. 18, 1920, *CPAE*, vol. 10, Calendar AEA, [35 372].

69. Einstein to Allgemeine Elektrizitäts-Gesellschaft Berlin, "Remarks to an Opinion Prepared for Mr. Sannig," Jan. 16, 1922, AEA, [35 378].

70. Georg Count von Arco to Einstein, Nov. 11, 1920, *CPAE*, vol. 10, doc. 199.

71. DE291604: Gesellschaft für drahtlose Telegraphie GmbH, "Einrichtung zur Erzeugung elektrischer Schwingungen," 1919.

72. DE310152: Dr. Erich F. Huth GmbH and Ludwig Kühn, "Schaltungsweise zur Schwingungserzeugung mit Vakuumröhren," 1919.

73. Albert Einstein, "Private Expert Opinion for Telefunken on the Patents of Meissner and Kühn," after Nov. 11, 1920, *CPAE*, vol. 7, doc. 48.

74. DE304283: Gesellschaft für drahtlose Telegraphie m. b. H., "Aus einer Kathodenstrahlröhre in Rückkopplungsschaltung bestehender Generator elektrischer Schwingungen," 1920. For details of both inventions, see Jonathan Zenneck, and Hans Rukop, *Lehrbuch der drahtlosen Telegraphie*, 5th ed. (Stuttgart: Enke, 1925), p. 613.

75. Albert Einstein, "Court Expert Opinion in the Matter of Signal Co. vs. Atlas Works," around Dec. 3, 1921, *CPAE*, vol. 7, doc. 66.

76. DE256747: Aurel Meckel, "Vorrichtung zur Bestimmung der Richtung von Schallwellen," 1913; DE257211: "Vorrichtung zur Bestimmung der Richtung von Schallwellen. Zusatz zum Patent 256747," 1913; DE257212: "Vorrichtung zur Bestimmung der Richtung von Schallwellen. Zusatz zum Patent 256747," 1913.

77. DE131235: Mario Russo D'Asar, "Vorrichtung zur Melden der Annäherung von Schiffen mittels unter Wasser angeordneter Schallaufnehmer," 1902.

78. GB1910-15102 (A): Thomas J. Bowlker, "Improvements in Apparatus for Submarine Signalling.," 1911; US224199: A. M. Mayer, "Topophone," 1879.

79. DE301669: Erich von Hornbostel, and Max Wertheimer, "Vorrichtung zur Bestimmung der Schallrichtung," 1920.

80. See *CPAE*, vol. 7, p. 478, n. 2.

81. Albert Einstein, "Court Expert Opinion in the Matter of Atlas Works vs. Signal Co.," Dec. 3, 1921, *CPAE*, vol. 7, doc. 67.

82. DE301669: Erich von Hornbostel, and Max Wertheimer, "Vorrichtung zur Bestimmung der Schallrichtung," 1920.

83. Albert Einstein, "Court Expert Opinion in the Matter of Signal Co. vs. Atlas Works," around Dec. 3, 1921, *CPAE*, vol. 7, doc. 66; US224199: A. M. Mayer, "Topophone," 1879; DE99667: David P. Heap, "Schallweiser mit zwei akustischen Empfängern," 1898; DE93144: E. Hardy, "Apparat zur Bestimmung der Herkunftsrichtung eines Schalles," 1897.

84. US964380: Thomas J. Bowlker, "Apparatus for Submarine Signaling," 1908. The same invention was patented in the United Kingdom as [81]; FR456318. Salomon, Charles. "Appareil destiné à déterminer la direction des ondes sonores et plus généralement de toutes ondes susceptibles d'être transformées en ondes sonores d'intensité proportionelle," 1916.

85. Heinrich Löwy, *Elektrodynamische Erforschung des Erdinneren und Luftschiffahrt* (Vienna: Manz, 1920).

86. Their opinions can be found in the Theodore von Kármán Collection, California Institute of Technology, Pasadena, California.

87. DE401448: Heinrich Löwy, "Verfahren zur Erforschung des Erdinnern mittels eines von einem Luftschiff oder Flugzeug über den Boden geführten offenen oder geschlossenen Schwingungskreises," 1924, submitted Nov. 2, 1921.

88. Heinrich Löwy to Theodore von Kármán, Nov. 16, 1921. He expressed his opinion in a similar vein in his letter to Willy Heilpern, Société Française des Pays Danubiens, written the same day. Both letters are in the Theodore von Kármán Collection, folder G7.1.

89. Albert Einstein, "Expert Opinion on Heinrich Löwy's Invention," Oct. 12, 1921, Theodore von Kármán Collection, folder G7.1.

90. Société Française des Pays Danubiens Marcel Schwob & Cie to Einstein, Nov. 7, 1921, *CPAE*, vol. 12, doc. 291.

91. Heinrich Löwy to Zeppelin Luftschiffbau GmbH, Mar. 9, 1922, Theodore von Kármán Collection, folder G7.1.

92. Rudolf Goldschmidt to Einstein, Mar. 7, 1921, *CPAE*, vol. 12, doc. 82.

93. E. Kreowski to Einstein, June 13, 1947, AEA, [35 537].

94. DE348111, 347785, 357007, and 370019.

95. US1386329: Rudolf Goldschmidt, "Mechanism for Converting Rotary into Reciprocatory Motion," 1921.

96. Rudolf Goldschmidt to Einstein, Jan. 14, 1922, AEA, [35 496].

97. US85721, 236697, 942299, 955339, 1091533, 1125500. 1192502, 1204245, 1249094, 1280269, 1286617, 1332864, 1363495, 1367117.

98. Expert Opinion on Goldschmidt's American Patent No. 1386329, no date [after Jan. 14, 1922], AEA, [35 494].

99. Paul Hausmeister to Einstein, Jan. 24, 1922, AEA, [43 865].

100. Einstein to Paul Hausmeister, Jan. 26, 1922, AEA, [43 867].

101. CH107196: Paul Hausmeister. "Verfahren zur Herstellung von Druckgasen," Oct. 1, 1924.

102. GB24260: Richard Eisenmann. "Improvements in or Relating to Electrically-Operated Musical Instruments," and A68245: Richard Eisenmann. "Umlaufender Stromunterbrecher zur Erzeugung von Tönen." Both were granted in 1914.

103. Richard Eisenmann to Einstein, July 18, 1922, AEA, [43 602].

104. US1350214: "Device for Regulating and Maintaining Constant the Speed of Motors," 1920.

105. Richard Eisenmann to Einstein, July 18, 1922, AEA, [43 602].

106. Einstein to Richard Eisenmann, July 27, 1922, AEA, [85 510].

107. Per Bloland, "The Electromagnetically-Prepared Piano and Its Compositional Implications," www.stanford.edu/~bloland/Assets/EMPP-Comp-Implications.pdf.

108. Max Gasser to Einstein, Feb. 4, 1948, AEA, [35 364]. He mentioned two patents (see notes 112 and 113 below).

109. Landgericht I to Einstein, Nov. 12, 1923, AEA, [35 361].

110. Einstein to Landgericht I, Apr. 4, 1923. He attached his opinion "Sachverständiges Gutachten zum Prozess Inag contra Optikon." It is available as a draft in Einstein's hand, AEA, [35 355], and as a signed typescript, [35 356].

111. DE298086: Firma Ed. Messter, "Verfahren zur Herstellung von photographischen Aufnahmen von Flugzeugen aus mittels einer Filmbandes," 1919; DE332233: "Verfahren zur photographischen Geländeaufnahne vom Flugzeug aus. Zusatz zum Patent 298086," 1921, or one of its relevant Austrian, Swiss, and Danish patents.

112. DE306384: Max Gasser, "Verfahren mittels dreier gegebener Punkte durch mechanische Ausmeßvorrichtungen mechanische Berechnungsapparate und durch geodätisch orientierte Doppelprojektionseinrichtungen lufttopographische Karten für eine photogeodätische Landesvermessung herzustellen," 1921. The process was also patented in the United States as US1585484: "Process and Apparatus for the Production of Aerogeodetical Stereophotographs," 1926.

113. DE306385: Max Gasser, "Verfahren und Vorrichtung zur Herstellung von Landkarten aus übergrefenden schiefen Aufnahmen," 1921.

114. Albert Einstein, "Gutachten zum Patentstreit der Deutschen Kabelwerke gegen Standard Telephones & Cables Ltd betreffend die Patente 341 678 und 390 178," Mar. 1928, AEA, [35 341].

115. DE341678: Western Electric Co., Inc, New York, "Verfahren zur Herstellung von Magnetkernen aus fein zerteilten, durch Isolierstoff voneinander getrennten magnetischen Stoffteilchen," 1921. Western Electric Co. had been the name of Standard Telephones before 1925.

116. DE390178: Bell Telephone Mfg. Co., Antwerpen, "Verfahren zur Herstellung von Magnetkernen für Pupin-Belastungsspulen, Magnete u. dgl.," 1924.

117. DE26813: C. Wetter, "Neuerung an Elektromagneten und Magnetkernen für dynamo-elektrische Maschinen und ähnliche Apparate," 1884.

118. DE226347: S. Hilpert, "Verfahren zur Herstellung magnetisierbarer Materialien von gleichzeitigen geringen elektrischer Leitfähigkeit für elektrische und magnetische Apparate," 1910; DE227788: "Zusatzpatent," 1910.

119. SU10133: I. E. Kechedzhiev-Kechedzhan (Кечеджиев-Кечеджан), "Описание трубы для наблюдения вблизи видимого наложения Солнца," Вестник Комитета по делам изобретений 5 (1929): 337. The source of name variants, description of the telescope, and administrative steps in the Soviet Union is Dieter Hoffmann's lecture published in D. Schulze, G. Wendel, K.-F. Wessel, and H. Scholz, eds., *Wissenschaftshistorisches Kolloquium anläßlich des 75. Geburtstages von Prof. em. Dr. habil. Friedrich Herneck* (Berlin: Sektion Wissenschaftstheorie und -organization der Humboldt-Universität zu Berlin, 1984), pp. 126–30.

120. Gesellschaft für Kulturelle Verbindung der Sowjetunion mit dem Auslande to Einstein, Feb. 19, 1930, AEA, [35 342].

121. Einstein to Gesellschaft für Kulturelle Verbindung der Sowjetunion mit dem Auslande, Feb. 25, 1930, AEA, [35 343].

122. Walter Kocherthaler to Einstein, June 15, 1934, AEA, [35 348].

123. Einstein to Walter Kocherthaler, June 24, 1934, AEA, [35 347].

124. US1951214: "Tinted Toilet Mirror" and US1951213: "Color-Filter Mirror" were granted to Peter Schlumbohm, Berlin, on Mar. 13, 1934.

125. Walter Kocherthaler to Einstein, July 5, 1934, AEA, [35 349].

126. Otto Henselman to Einstein, Aug. 7, 1944, AEA, [55 279].

127. US2215701: Otto Henselman, "Bearing Roller," 1940.

128. Einstein to Otto Henselman, Sept. 13, 1944, AEA, [55 283]. A German draft is also available, AEA, [55 280].

CHAPTER 4: European Inventions

1. Albert Einstein, "Eine neue elektrostatische Methode zur Messung kleiner Elektrizitätsmengen," *Physikalische Zeitschrift* 9 (1908): 216–17, *Collected Papers of Albert Einstein* (*CPAE*; Princeton, NJ: Princeton University Press, 1987–2009), vol. 2, doc. 48.

2. Albert Einstein, "Zur Theorie der Brownschen Bewegung," *Annalen der Physik* 19 (1906): 371–81; "Theoretische Bemerkungen über die Brownsche Bewegung," *Zeitschrift für Elektrochemie und angewandte physikalische Chemie* 13 (1907): 41–42; "Ueber die Natur der Bewegungen mikroskopisch kleiner, in Flüssigkeiten suspendierter Teilchen," *Naturforschende Gesellschaft Bern. Mitteilungen* (1907): no. 1038; and "Elementare Theorie der Brownschen Bewegung," *Zeitschrift für Elektrochemie und angewandte physikalische Chemie* 14 (1908): 235–239, *CPAE* vol. 2, docs. 32, 40, 43, and 50, resp.

3. Albert Einstein, "Über die Gültigkeitsgrenze des Satzes vom thermodynamischen Gleichgewicht und über die Möglichkeit einer neuen Bestimmung der Elementarquanta," *Annalen der Physik* 22 (1907): 569–72, *CPAE*, vol. 2, doc. 39.

4. Einstein to Conrad and Paul Habicht, July 15, 1907, *CPAE*, vol. 5, doc. 48.

5. Einstein to Paul and Conrad Habicht, Aug. 16, 1907, *CPAE*, vol. 5, doc. 54.

6. Einstein to Conrad and Paul Habicht, Sept. 2, 1907, *CPAE*, vol. 5, doc. 56.

7. Einstein to Conrad Habicht, Dec. 24, 1907, *CPAE*, vol. 5, doc. 69.

8. Emil Bose to Einstein, Feb. 12, 1908, *CPAE*, vol. 5, doc. 83.

9. Paul Habicht to Einstein, Feb. 19, 1908, *CPAE*, vol. 5, doc. 86.

10. Adolf Gasser to Einstein, Mar. 9, 1908, *CPAE*, vol. 5, doc. 92.

11. Albert Einstein, "Eine neue elektrostatische Methode zur Messung kleiner Elektrizitätsmenge," *Physikalische Zeitschrift* 9 (1908): 216–17, *CPAE*, vol. 2, doc. 48.

12. Joseph Kowalski to Einstein, Mar. 30, 1908, *CPAE*, vol. 5, doc. 94.

13. Paul Habicht to Einstein, June 1908, *CPAE*, vol. 5, doc. 104.

14. Paul Habicht to Einstein, Apr. 4, 1908, *CPAE*, vol. 5, doc. 95.

15. Paul Habicht to Einstein, May 17, 1908, *CPAE*, vol. 5, doc. 99.

16. See Paul Habicht to Einstein, Apr. 4, 1908, *CPAE*, vol. 5, doc. 95, n. 5.

17. Einstein to Jakob Laub, after Nov. 1, 1908, *CPAE*, vol. 5, doc. 125.

18. Einstein to Albert Gockel, Dec. 3, 1908, *CPAE*, vol. 5, doc. 130.

19. Einstein to Jakob Laub, Mar. 20, 1909, *CPAE*, vol. 5, doc. 143.

20. Einstein to Albert Gockel, Mar. 25(?), 1909, *CPAE*, vol. 5, doc. 144.

21. Einstein to Conrad Habicht, Apr. 15, 1909, *CPAE*, vol. 5, doc. 150.

22. Einstein to Conrad Habicht, Sept. 3, 1909, *CPAE*, vol. 5, doc. 177.

23. Einstein to Conrad Habicht, Nov. 5, 1909, *CPAE*, vol. 5, doc. 185.

24. Einstein to Conrad Habicht, Mar. 4, 1910, *CPAE*, vol. 5, doc. 198.

25. Conrad Habicht, and Paul Habicht, "Elektrostatischer Potentialmultiplikator nach A. Einstein," *Physikalische Zeitschrift* 11 (1910): 532–35.

26. Einstein to Heinrich Zangger, Dec. 25, 1911, *CPAE*, vol. 5, doc. 330.

27. Einstein to Michele Besso, Feb. 4, 1912, *CPAE*, vol. 5, doc. 354.

28. Paul Habicht to Einstein, Dec. 27, 1911, *CPAE*, vol. 5, doc. 332.

29. Conrad Habicht, and Paul Habicht, "Essai de démonstration avec le multiplicateur de potentiel d'après Einstein," *Archives des sciences physiques et naturelles* 33 (1912): 258–59. See also Einstein to Michele Besso, Mar. 26, 1912, *CPAE*, vol. 5, doc. 377.

30. For a comprehensive account of the development of the *Maschinchen*, see the editorial note "Einstein's 'Maschinchen' for the Measurement of Small Quantities of Electricity," in *CPAE*, vol. 5, pp. 51–55.

31. Albert Einstein, "Méthode pour la détermination de valeurs statistiques d'observations concernant des grandeurs soumises des fluctuations irrégulières," *Archives des sciences physiques et naturelles* 37 (1914): 254–56, *CPAE*, vol. 4, doc. 29.

32. Albert Einstein, "Eine Methode zur statistischen Verwertung von Beobachtungen scheinbar unregelmässig quasiperiodisch verlaufender Vorgänge," [after 28 Feb. 1914], *CPAE*, vol. 4, doc. 30.

33. Einstein to Adolf Schmidt, Oct. 30, 1914, *CPAE*, vol. 8, doc. 37.

34. Adolf Schmidt, "Ein Planimeter zur Bestimmung der mittleren Ordinaten beliebiger Abschnitte von registrierten Kurven," *Zeitschrift für Instrumentenkunde* 25 (1905): 261–73.

35. Adolf Schmidt to Einstein, Oct. 31, 1914, *CPAE*, vol. 8, doc. 38.

36. Albert Einstein, "Elementare Theorie der Wasserwellen und des Fluges," *Die Naturwissenschaften* 4 (1916): 509–10, *CPAE*, vol. 6, doc. 39.

37. N. Joukowsky, "Über die Konturen der Tragflächen der Drachenflieger," *Zeitschrift für Flugtechnik und Motorluftschiffahrt* 1 (1910): 281–84.

38. Ludwig Prandtl, "Abriss der Lehre von der Flüssigkeits- und Gasbewegung," in *Handwörterbuch der Naturwissenschaften* (Jena: Fischer, 1913), pp. 101–40.

39. M. Wilhelm Kutta, "Über eine mit der Grundlagen des Flugsproblems in Beziehung stehende zweidimensionale Strömung," *Königlich-Bayerische Akademie der Wissenschaften* (Munich), *Sitzungsberichte. Mathematisch-physikalische Klasse* (1910): 1–58; "Über ebene Zirkulationsströmungen nebst flugtechnischen Anwendungen," *Königlich-Bayerische Akademie der Wissenschaften* (Munich), *Sitzungsberichte. Mathematisch-physikalische Klasse* (1911): 65–125.

40. Ludwig Prandtl, and Frederick Lanchester, "Tragflügeltheorie. I, II," *Gesellschaft der Wissenschaften zu Göttingen. Nachrichten* (1918): 451–77; (1919): 107–37.

41. Paul Ehrhardt to Einstein, Aug. 26, 1954, Albert Einstein Archives, Jerusalem (AEA), [59 556].

42. Carl Seelig, *Albert Einstein. Leben und Werk eines Genies unserer Zeit* (Zurich: Europa, 1960), p. 251.

43. *W. W. I Aero*, no. 118 (Feb. 1988): 43.

44. Albrecht Fölsing, *Albert Einstein: Eine Biographie*, 2nd ed. (Frankfurt am Main: Suhrkamp, 1993), pp. 446–47; English translation, pp. 399–400.

45. Paul Ehrhardt to Einstein, Sept. 26, 1954, AEA, [59 557].

46. Einstein to Michele Besso, May 14, 1916, *CPAE*, vol. 8, doc. 219.

47. See Michael Eckert, *The Dawn of Fluid Dynamics: A Discipline between Science and Technology* (Weinheim, Germany: Wiley-VCH, 2006), p. 68.

48. Julius C. Rotta, *Die Aerodynamische Versuchsanstalt in Göttingen. Ein Werk Ludwig Prandtls, ihre Geschichte von den Anfangen bis 1925* (Göttingen, Germany: Vanderhoeck & Ruprecht, 1990), p. 174.

49. See Michael Eckert, *The Dawn of Fluid Dynamics: A Discipline between Science and Technology* (Weinheim, Germany: Wiley-VCH, 2006), p. 61.

50. Max Munk, and Carl Pohlhausen, "Messungen an einfachen Flügelprofilen," *Technische Berichte* 1 (1917): table 149.

51. Ibid.

52. Peter Grosz, "Herr Dr. Prof. Albert Who? Einstein the Aerodynamicist That's Who! Or Albert Einstein and His Role in German Aviation in World War I," *W. W. I Aero*, no. 118 (Feb. 1988): 42–46.

53. Paul Ehrhardt to Einstein, Aug. 26, 1954, AEA, [59 556].

54. Einstein to Paul Ehrhardt, Sept. 7, 1954, AEA, [76 235].

55. Fragment of a letter of Einstein to Otto Marx, Dec. 22, 1917, *CPAE*, vol. 8, doc. 415.

56. Grosz, "Herr Dr. Prof. Albert Who?"

57. Seelig, *Albert Einstein. Leben und Werk eines Genies unserer Zeit*, p. 252.

58. *Österreichisches Aero Club. Mitteilungen* (1920), p. 140.

59. Hans Usener, *Der Kreisel als Richtungsweiser. Seine Entwickelung. Theorie und Eigenschaften* (Munich: Militärische Verlagsanstalt, 1917), sent to Einstein by Anschütz and also by Usener himself, *CPAE*, vol. 8, docs. 603 and 606.

60. Hermann Anschütz-Kaempfe to Einstein, Oct. 10, 1920, *CPAE*, vol. 10, doc. 172.

61. Hermann Anschütz-Kaempfe to Einstein, Dec. 28, 1920, *CPAE*, vol. 10, doc. 247.

62. Hermann Anschütz-Kaempfe to Einstein, Mar. 10, 1921, *CPAE*, vol. 12, doc. 92.

63. Einstein to Hermann Anschütz-Kaempfe, Mar. 13, 1921, *CPAE*, vol. 12, doc. 95.

64. Hermann Anschütz-Kaempfe to Einstein, July 23, 1921, *CPAE*, vol. 12, doc. 191.

65. Einstein to Hermann Anschütz-Kaempfe, July 28, 1921, *CPAE*, vol. 12, doc. 194.

66. Schleswig-Holsteinische Landesbibliothek, Kiel, Zg.-Nr.: 57/1992, *CPAE*, vol. 12, appendix A.

67. Einstein to Hermann Anschütz-Kaempfe, Sept. 18, 1921, *CPAE*, vol. 12, doc. 239.

68. Hermann Anschütz-Kaempfe to Einstein, Sept. 20, 1921, *CPAE*, vol. 12, doc. 241.

69. Einstein to Hermann Anschütz-Kaempfe, Sept. 17, 1921, *CPAE*, vol. 12, doc. 237.

70. Einstein to Hermann Anschütz-Kaempfe, Sept. 18, 1921, *CPAE*, vol. 12, doc. 239.

71. Einstein to Hermann Anschütz-Kaempfe, Mar. 13, 1921, *CPAE*, vol. 12, doc. 95.

72. Einstein to Hermann Anschütz-Kaempfe, June 18, 1922, AEA, [80 280].

73. Hermann Anschütz-Kaempfe to Einstein, Aug. 20, 1922, AEA, [37 382].

74. Hermann Anschütz-Kaempfe to Einstein, Dec. 6, 1923, AEA, [37 389].

75. Hermann Anschütz-Kaempfe to Einstein, Nov. 9, 1924, in Dieter Lohmeier and Bernhardt Schell, eds., *Einstein, Anschütz und der Kieler Kreiselkompass–Einstein, Anschütz and the Kiel Gyro Compass*, 2d ed. (Kiel, Germany: Raytheon Marine GmbH, 2005), no. 57.

76. Einstein to Hermann Anschütz-Kaempfe, Aug. 31, 1925, AEA, [37 395].

77. "Giro" to Einstein, Oct. 11, 1926, AEA, [35 401].

78. Einstein to Wolfgang Otto, Oct. 28, 1926, AEA, [80 660].

79. Wolfgang Otto to Einstein, Nov. 2, 1926, AEA, [37 401].

80. Hermann Anschütz-Kaempfe to Einstein, June 25 and July 2, 1922, AEA, [37 398] and [37 379].

81. Einstein to Elsa Einstein, Apr. 28, 1923, AEA, [143 129].

82. Einstein to Hermann Anschütz-Kaempfe, Sept. 8(?), 1923, AEA, [80 282].

83. Einstein to Karl Glitscher, May 27, 1925, AEA, [37 394].

84. Hermann Anschütz-Kaempfe to Arnold Sommerfeld, July 12, 1922, in Lohmeier and Schell, *Einstein, Anschütz und der Kieler Kreiselkompass*, no. 37.

85. Hermann Anschütz-Kaempfe to Reta Anschütz, Nov. 2, 1920, Deutsches Museum, München, Nachlass Anschütz.

86. Karl Glitscher to Hermann Anschütz-Kaempfe, Mar. 31, 1921, in Lohmeier and Schell, *Einstein, Anschütz und der Kieler Kreiselkompass*, p. 40.

87. Einstein to Elsa Einstein, Apr. 28, 1923, AEA, [143 129].

88. Max Schuler, "Die geschichtliche Entwicklung des Kreiselkompasses in Deutschland. Teil 1: Schiffskreiselkompasse," *VDI Zeitschrift* 104 (1962): 469–76.

89. Einstein to Arnold Sommerfeld, Feb. 5, 1919, *CPAE*, vol. 9, doc. 5.

90. Einstein to Elsa Einstein, Sept. 14, 1920, *CPAE*, vol. 10, doc. 149.

91. Hermann Anschütz-Kaempfe to Einstein, Oct. 10, 1920, *CPAE*, vol. 10, doc. 172.

92. Einstein to Elsa Einstein, Sept. 14, 1920, *CPAE*, vol. 10, doc. 149.

93. Einstein to Elsa Einstein, Apr. 28, 1923, AEA, [143 129].

94. Einstein to Elsa Einstein, Sept. 14, 1920, *CPAE*, vol. 10, doc. 149.

95. Hermann Anschütz-Kaempfe to Einstein, Dec. 19, 1920, *CPAE*, vol. 10, doc. 237.

96. Hermann Anschütz-Kaempfe to Einstein, Nov. 10, 1921, *CPAE*, vol. 12, doc. 293.

97. Einstein to Arnold Sommerfeld, Dec. 18–20, 1920, *CPAE*, vol. 10, doc. 236.

98. Hermann Anschütz-Kaempfe to Einstein, Dec. 28, 1920, *CPAE*, vol. 10, doc. 247.

99. Einstein to Hermann Anschütz-Kaempfe, July 22, 1921, *CPAE*, vol. 12, doc. 189.

100. Einstein to Mileva Einstein-Marić, Aug. 21, 1921, *CPAE*, vol. 12, doc. 218.

101. Einstein to Hermann Anschütz-Kaempfe, July 1, 1922, AEA, [80 288].

102. Albert Einstein, "In memoriam Walther Rathenau," *Neue Rundschau* 33, part 2 (1922): 815–16.

103. Einstein to Max Planck, July 6, 1922, AEA, [77 023].

104. Einstein to Marie Curie, July 11, 1922, AEA, [34 776].

105. Hermann Anschütz-Kaempfe to Arnold Sommerfeld, July 12, 1922, in Lohmeier and Schell, *Einstein, Anschütz und der Kieler Kreiselkompass*, no. 37.

106. Einstein to Hermann Anschütz-Kaempfe, July 12, 1922, AEA, [80 721].

107. Ibid.

108. Einstein to Hermann Anschütz-Kaempfe, July 16, 1922, AEA, [80 720].

109. Hermann Anschütz-Kaempfe to Einstein, July 19, 1922, AEA, [37 381].

110. Hermann Anschütz-Kaempfe to Einstein, Mar. 31, 1923, AEA, [37 383].

111. Hermann Anschütz-Kaempfe to Einstein, Dec. 28, 1920, *CPAE*, vol. 10, doc. 247.

112. Hermann Anschütz-Kaempfe to Einstein, Mar. 18, 1922, AEA, [37 375].

113. Hermann Anschütz-Kaempfe to Einstein, June 9, 1922, AEA, [37 377].

114. Einstein to Elsa Einstein, Apr. 21, [1923], AEA, [143 128].

115. "Giro" to Einstein, Oct. 11, 1926, AEA, [35 401.0].

116. DE394667: Anschütz & Co., "Kreiselapparat für Meßzwecke," 1924.

117. Einstein to Hermann Anschütz-Kaempfe, Feb. 17, 1925, AEA, [84 235].

118. Albert Einstein and Hans Mühsam, "Experimentelle Bestimmung der Kanalweite von Filtern," *Deutsche medizinische Wochenschrift* 4 (1923): 1012–13.

119. Hans Mühsam, "Zur Eichung von Filters," *Gesundheits-Ingenieur* 46 (1923): 440.

120. Karol J. Mysels, "Einstein's Last Contribution to Surface Chemistry," *Langmuir* 5 (1989): 1265–67.

121. *Standard Test Method for Maximum Pore Diameter and Permeability of Rigid Porous Filters for Laboratory Use*, ASTM E128-99 (2005).

122. Einstein to Michele Besso, Dec. 12, 1919, *CPAE*, vol. 9, doc. 207; Einstein to Paul Ehrenfest, Sept. 1, 1921, *CPAE*, vol. 12, doc. 219.

123. Walther Nernst to Einstein, July 29, 1921, *CPAE*, vol. 12, doc. 195.

124. Karl Wolfgang Graff, "Albert Einstein als Erfinder in den Jahren 1907–1933" (Thesis, Historisches Institut der Universität Stuttgart, 2004), p. 213.

125. Contract proposal to A. Borsig GmbH, Berlin-Tegel, Feb. 1922, AEA, [18 447], [35 352].

126. Einstein to the board of trustees of the Kaiser Wilhelm Institute of Physics, Jan. 28, 1922, AEA, [81 934].

127. Einstein to Hans Albert and Eduard Einstein, Mar. 4, 1922, AEA, [75 660].

128. Graff, "Albert Einstein als Erfinder in den Jahren 1907–1933," p. 217.

129. Hans Albert Einstein to Einstein, between Feb. 12 and Mar. 4, 1922, AEA, [144 029].

130. For circumstances and possible details, see Graff, "Albert Einstein als Erfinder in den Jahren 1907–1933," pp. 213–17.

131. Leó Szilárd to Einstein, Sept. 10, 1926, AEA, [35 555].

132. Leó Szilárd to H. Peiser (Bamag-Meguin), Nov. 8, 1926, and "Vertragentwurf" attached to it, AEA, [35 562], [35 563].

133. Leó Szilárd to Béla Szilárd, Oct. 26, 1926, AEA, [35 560].

134. Graff, "Albert Einstein als Erfinder in den Jahren 1907–1933," p. 227.

135. Einstein to Leó Szilárd, Sept. 15, 1928, AEA, [21 432]. The card was not mailed.

136. DE456152: Platen-Munters Refrigerating System Aktiebolag, "Verfahren zur Steigerung des Umlaufs eines Hilfsmittels in Absorptionskälteapparaten," 1928.

137. DE410715: Platen-Munters Refrigerating System Aktiebolag, "Verfahren zur Kälteerzeugung nach dem Absorptions-Diffusionsprinzip," 1925. Patented in Sweden in 1911.

138. Chris Holland, "We Are Going Back to Future," www.democraticunderground .com/discuss/duboard.php?az=view_all&address=115x142895. I thank Alice Calaprice for turning my attention to this report.

139. Andy Delano, "Design Analysis of the Einstein Refrigeration Cycle" (Ph.D. dissertation, Georgia Institute of Technology, June 1998), www.me.gatech.edu/energy /andy_phd/.

140. DE441752: Clemens Bergl and Walther Dietrich, "Verfahren zur Erzeugung von Kälte mit Hilfe organischer Flüssigkeiten ohne Wiedergewinnung derselben als solcher," 1927.

141. Graff, "Albert Einstein als Erfinder in den Jahren 1907–1933," p. 263.

142. Ibid., pp. 260–63.

143. Leó Szilárd to Einstein, Mar. 22, 1930, AEA, [35 586].

144. DE533945: Leó Szilárd, "Pumpe," 1931; DE531581: Leó Szilárd, "Pumpe, insbesondere zur Verdichtung von Gasen und Dämpfen in Kältemaschinen," 1933.

145. DE548136: Leó Szilárd, "Kältemaschine," 1932.

146. DE570959: Leó Szilárd, "Vorrichtung zur Bewegung von flüssigem Metall," 1933.

147. DE556536: Leó Szilárd, "Kältemaschine," 1932.

148. DE533945: Leó Szilárd, "Pumpe," 1931.

149. DE564680: Leó Szilárd, "Kältemaschine," 1932.

150. DE562898: Leó Szilárd, "Wärmeübertrager," 1932.

151. DE581780: Leó Szilárd, "Kompressor, insbesondere für Kältemaschinen," 1933.

152. DE562523: Leó Szilárd, "Absperrorgan," 1932.

153. DE543214: Leó Szilárd, "Vorrichtung zur Bewegung von flüssigen Metallen," 1932; DE555141: "Vorrichtung zur Bewegung von flüssigen Metallen, 1932; DE568680: "Stator für Kältemaschinen," 1933.

154. [Albert Korodi and László Bihaly], AEG Technische Bericht No. 689: Entwicklungsarbeiten an einer Kompressions-Kältemaschine mit Wanderfeld-Flüssigkeitmotor, Aug. 16, 1932, Deutsches Technikmuseum, Berlin.

155. Einstein to Leó Szilárd, Sept. 15, 1928, AEA, [21 432]. The card was not mailed.

156. DE476812: Leó Szilárd, "Verfahren zum Gießen von Metallen in Formen unter Anwendung elektrischer Ströme," 1929. Application date Jan. 20, 1926.

157. US853789: Frank Holden, "Mercury-Meter," 1907.

158. Graff, "Albert Einstein als Erfinder in den Jahren 1907–1933," p. 274.

159. DE319231 and GB126947: Julius F. G. P. Hartmann, "Improvements in or Relating to Apparatus for Producing a Continuous Electrically Conducting Liquid Jet," 1919; DE281727: Brown, Boveri & Cie, "Verfahren und Einrichtung zur Herstellung eines Vakuums oder einer Verdichtung von Gasen oder Dämpfen," 1914; and Technical Report of GEC, no. 17,351. See Graff, "Albert Einstein als Erfinder in den Jahren 1907–1933," p. 274.

160. DE511137: Allgemeine Elektrizitäts-Gesellschaft (Waldemar Brückel), "Elektrodynamische Pumpe," 1930.

161. US1792449: Millard C. Spencer, "Fluid-Conductor Motor," 1931.

162. US1660407: Kenneth T. Bainbridge, "Liquid-Conductor Pump," 1928.

163. Leó Szilárd to Einstein, Sept. 27, 1930, AEA, [35 590].

164. Leó Szilárd to Einstein, Oct. 10, 1931, AEA, [21 437].

165. Leó Szilárd to Einstein, July 27, 1932, AEA, [35 618].

166. Leó Szilárd to Einstein, Sept. 27, 1930, AEA, [35 590].

167. Leó Szilárd to Einstein, Mar. 22, 1930, AEA, [35 586].

168. Leó Szilárd to Einstein, July 23, 1931, AEA, [35 600].

169. Gano Dunn to Einstein, Feb. 21 and 25, 1931, AEA, [35 591] and [35 593], resp.

170. "Finds New Method for Refrigeration. D. Daniel F. Comstock, Research Engineer, Invents Method Adapted to Cool Homes," *Wall Street Journal*, Aug. 22, 1930.

171. US1924914: Daniel F. Comstock, "Absorption System," 1930.

172. Leó Szilárd to Einstein, Apr. 3, 1931, AEA, [35 597].

173. Patent Rights Transfer Request by Szilard, Dec. 1, 1931, AEA, [35 617].

174. Einstein to Leó Szilárd, July 6, 1932, AEA, [35 614].

175. Leó Szilárd to Einstein, July 14, 1932, AEA, [35 616].

176. Leó Szilárd to Einstein, Sept. 10, 1926, AEA, [35 555].

177. Recommendation at the bottom of Leó Szilárd to Einstein, June 30, 1931, AEA, [35 598].

178. Einstein to American General Consulate, Oct. 27, 1931, AEA, [21 440].

179. Reichspatentamt to Einstein, Mar. 20, 1933, AEA, [35 623].

180. Reichspatentamt to Einstein, Apr. 27, 1933, AEA, [35 624], [35 625].

181. Melanie Jaeger to Einstein, Apr. 9, 1934, AEA, [35 626]; Einstein to Melanie Jaeger, Apr. 13, 1934, AEA, [35 627]; Melanie Jaeger to Einstein, n.d., AEA, [35 629]; Melanie Jaeger to Einstein, n.d., AEA, [35 630].

182. Julius Janowitz to Einstein, Apr. 26, 1934, AEA, [35 628].

183. Leó Szilárd to Einstein, Apr. 1, 1927, AEA, [35 567].

184. Leó Szilárd to Einstein, Oct. 12, 1929, AEA, [35 585].

185. Leó Szilárd to Einstein, Oct. 12, 1929, AEA, [35 584].

186. H. R. G. Goldie to Otto Nathan, Aug. 3, 1975, AEA, [35 540].

187. DE590783: Albert Einstein and Rudolf Goldschmidt, "Vorrichtung, insbesondere für Schallwiedergabegeräte, bei der elektrische Stromänderungen durch Magnetostriktion Bewegungen eines Magnetkörpers hervorrufen," 1934.

188. Rudolf Goldschmidt to Einstein, Feb. 28, 1929, AEA, [35 501].

189. Einstein to Rudolf Goldschmidt, Mar. 8, 1929, AEA, [35 502].

190. "Magnetische Einrichtung, insbesondere zur Betätigung von Lautsprechern, Relais und Messinstrumenten," AEA, [35 506].

191. DE590783: Albert Einstein and Rudolf Goldschmidt, "Vorrichtung, insbesondere für Schallwiedergabegeräte, bei der elektrische Stromänderungen durch Magnetostriktion Bewegungen eines Magnetkörpers hervorrufen," 1934.

192. "Elektromagnetische Antriebsystem für Lautsprecher," AEA, [35 504]; "Elektromagnetisches Triebsystem," AEA, [35 505].

193. E. Kreowski to Einstein, June 13, 1947, AEA, [35 537].

194. Tekniske Forsøgsaktieselskab to Einstein, Apr. 9, 1929, AEA, [35 507].

195. DE521989: Rudolf Goldschmidt, "Verfahren zum Einregulierung von Telephonmembranen nach dem Zusammenbau mit dem Magneten," 1931.

196. Einstein to Rudolf Goldschmidt, Apr. 12, 1929, AEA, [35 510].

197. Tekniske Forsøgsaktieselskab to Einstein, Apr. 9, 1929, AEA, [35 507].

198. AEA-Pix [64 022]. Translated by Jane Dietrich.

199. Rudolf Goldschmidt to Einstein, Nov. 30, 1928, AEA, [35 498]. Translated by Jane Dietrich.

200. Rudolf Goldschmidt to Einstein, Sept. 25, 1929, AEA, [35 512].

201. Rudolf Goldschmidt to Einstein, May 2, 1928, AEA, [35 499].

202. GB321395: Rudolf Goldschmidt, "Diaphragm Especially for the Reproduction of Sound," 1929.

203. Rudolf Goldschmidt to Einstein, Nov. 15, 1928, AEA, [35 500].

204. Rudolf Goldschmidt to Einstein, Oct. 31, 1931, AEA, [35 513].

205. Rudolf Goldschmidt to Einstein, Nov. 2, 1932, AEA, [35 514].

206. Bruno Eisner, "Begegnung mit Einstein," *Aufbau* (*Reconstruction,* an American weekly published in New York), Beiblatt "Der Zeitgeist," Mar. 9, 1962, pp. 21–22 (abbreviated publication of pp. 129–38 of his remembrances), AEA, [81 102].

207. Rudolf Goldschmidt to Einstein, Nov. 6, 1932, AEA, [35 515].

208. Rudolf Goldschmidt to Einstein, Nov. 13, 1932, AEA, [35 516].

209. Mrs. Mendelssohn to Einstein, Nov. 21, 1932, AEA, [38 517].

210. Rudolf Goldschmidt to Einstein, Nov. 26, 1932, AEA, [35 518], [35 519].

211. Rudolf Goldschmidt to Einstein, Aug. 19, 1933, AEA, [35 520].

212. Rudolf Goldschmidt to Einstein, Aug. 22 and Aug. 24, 1933, AEA, [35 521] and [35 522].

213. Eisner, "Begegnung mit Einstein."

214. Rudolf Goldschmidt to Einstein, Oct. 31, 1941, AEA, [35 523].

215. Einstein to Rudolf Goldschmidt, Dec. 20, 1941, AEA, [35 527].

216. GB553955: Rudolf Goldschmidt, "Improvements in or Relating to Electro-Magnetic Sound-Transmission Apparatus," 1943.

217. Otto Nathan to H. R. G. Goldie, Oct. 25, 1957, AEA, [35 539].

218. H. R. G. Goldie to Otto Nathan, Aug. 3, 1975, AEA, [35 540].

219. Eisner, "Begegnung mit Einstein."

220. H. R. G. Goldie to Otto Nathan, Aug. 3, 1975, AEA, [35 540].

221. Rudolf Goldschmidt to Einstein, Nov. 6, 1932, AEA, [35 515].

222. Rudolf Goldschmidt to Einstein, Aug. 22 and Aug. 24, 1933, AEA, [35 521] and [35 522].

223. Rudolf Goldschmidt to Einstein, Oct. 5 and Nov. 14, 1942, AEA, [35 529] and [35 534].

CHAPTER 5: American Inventions

1. For Bucky's life and work, see Karin Bormacher, "Gustav Bucky (1880–1963) Biobibliographie eines Röntgenologen und Erfinders" (Inaugural-Dissertation, Freie Universität, Berlin, 1967).

2. Einstein to Gustav Bucky, [Nov. 8, 1934], Albert Einstein Archives, Jerusalem (AEA), [35 417].

3. Einstein to Gustav Bucky, Nov. 12, 1934, AEA, [35 416].

4. Gustav Bucky to Einstein, Jan. 25, 1935, AEA, [35 422].

5. "Liste der zu bearbeitenden Ideen," [Jan. 1935], AEA, [35 423].

6. Gustav Bucky to Einstein, Feb. 10, 1935, AEA, [35 424].

7. Einstein to Gustav Bucky, Feb. 11, 1935, AEA, [35 426].

8. Gustav Bucky to Emil Mayer, Mar. 5, 1935, AEA, [35 452].

9. Gustav Bucky to Emil Mayer, Mar. 5, 1935: "Gasdurchlaessiges nicht benetzbares Gewebe," AEA, [35 455].

10. Emil Mayer to Gustav Bucky, Mar. 15, 1935, AEA, [35 444].

11. Gustav Bucky to Emil Mayer, Mar. 16, 1935, AEA, [35 443].

12. Emil Mayer to Gustav Bucky, June 5, 1935, AEA, [35 429]. It was filed on June 6, 1935, under serial number 25,239.

13. Attachment to Emil Mayer's letter to Gustav Bucky, June 5, 1935, AEA, [35 439].

14. Gustav Bucky to Einstein, June 7, 1935, AEA, [35 427], [35 428].

15. Emil Mayer to Gustav Bucky, June 5, 1935, AEA, [35 429]. It was filed on June 6, 1935, under serial number 25,239.

16. Gustav Bucky to Emil Mayer, Mar. 5, 1935: "Billiges waermeisolierendes Verpackungsgefaess," AEA, [35 454].

17. Josef Oppenheimer to Einstein [?], Apr. 18, 1936, AEA, [35 439.1], [35 440].

18. Gustav Bucky to Emil Mayer, Mar. 5, 1935: "Fluessigkeitsreinigung durch starke elektrische Felder," AEA, [35 454.1].

19. Emil Mayer to Gustav Bucky, Mar. 15, 1935, AEA, [35 444].

20. Gustav Bucky to Emil Mayer, Mar. 16, 1935, AEA, [35 443].

21. Briesen & Schrenk to Einstein, July 10, 1935, AEA, [35 456.1].

22. US1743526: Charles T. Cabrera, "Filtering Medium," 1930.

23. Briesen & Schrenk to Einstein, July 10, 1935, AEA, [35 457].

24. Einstein to Briesen & Schrenk, July 13, 1935, AEA, [35 462]. A typed file copy [35 461] bears the date July 15.

25. Walter S. Bleistein to Bucky, Nov. 28, 1934, AEA, [35 418].

26. "Abkommen zwischen der Roefinag, Aktiengesellschaft . . . und den Herren Dr. Gustav Bucky . . . und Professor Dr. Albert Einstein," [after Mar. 5, 1935], AEA, [35 420].

27. Roefinag, Aktiengesellschaft in Zürich Schweiz, "Vorrichtung zum selbsttätigen Einbeziehen eines Korrekturfaktors in die Anzeigen eines Messinstrumentes," [Nov. 28, 1934], AEA, [35 421]); Gustav Bucky to Emil Mayer, Mar. 5, 1935: "Vorrichtung zur automatische Korrektionsablesung," AEA, [35 453.1].

28. Emil Mayer to Einstein, Mar. 12, 1935, AEA, [35 442]. Einstein's manuscript on the gyro apparatus is dated January 1935, and its copy, typed on February 28, 1935 [35 441], is attached to Mayer's letter.

29. Briesen & Schrenk to Einstein, July 10 and 11, 1935, AEA, [35 456], [35 458], [35 459].

30. GB19073785: John T. Morrison, "Improvements in Gyrostatic Apparatus," 1908.

31. Einstein to Briesen & Schrenk, July 13, 1935, AEA, [35 462]. A typed file copy [35 461] bears the date July 15.

32. Einstein to Gustav Bucky, July 20, 1935, AEA, [35 434].

33. Einstein to Gustav Bucky, Oct. 11, 1940, AEA, [35 446]. Its English translation is also available, AEA, [35 447].

34. Walter S. Bleston to Einstein, Nov. 1, 1940, AEA, [35 445].

35. Herbert Thompson to Einstein, May 6, 1941, AEA, [35 449].

36. Einstein to Gustav Bucky, no date, AEA, [35 450].

37. Einstein to Gustav Bucky, [Sept. 21, 1942] Monday, AEA, [35 482].

38. Gustav Bucky to Einstein, June 7, 1935, AEA, [35 427].

39. Briesen & Schrenk to Einstein, July 10, 1935, AEA, [35 456.1].

40. US1783138: Aage V. Petersen, "Transforming of Acoustic Swingings into Electric Capacity Swingings," 1930.

41. US1687231: James B. Speed, "Translating Device," 1928.

42. Einstein to Briesen & Schrenk, July 13, 1935, AEA, [35 462]. A typed file copy [35 461] bears the date July 15.

43. Briesen & Schrenk to Gustav Bucky, July 18, 1935, AEA, [35 464].

44. At the New Brunswick station of RCA. József Illy, ed., *Albert Meets America: How Journalists Treated Genius during Einstein's 1921 Travels* (Baltimore: Johns Hopkins University Press, 2006), p. 130.

45. Einstein to Gustav Bucky, July 13, 1935, AEA, [35 460].

46. Einstein to Gustav Bucky, July 20, 1935, AEA, [35 434].

47. David Sarnoff to Einstein, July 22, 1935, AEA, [51 590].

48. Einstein to Gustav Bucky, July 24, 1935, AEA, [35 466].

49. M. C. Batsel to Einstein, Aug. 14, 1935, AEA, [25 368], [25 369].

50. Elsworth D. Cook to M. C. Batsel, Aug. 13, 1935, AEA, [25 369].

51. Briesen & Schrenk to Einstein, Aug. 20, 1935, AEA, [35 467].

52. Einstein to M. C. Batsel, Aug. 22, 1935, AEA, [25 370].

53. Einstein to Briesen & Schrenk, Aug. 23, 1935, AEA, [35 468].

54. Einstein to M. C. Batsel, Aug. 26, 1935, AEA, [25 371].

55. M. C. Batsel to Einstein, Sept. 5, 1935, AEA, [25 373].

56. Einstein to M. C. Batsel, Sept. 6, 1935, AEA, [25 374].

57. Einstein to M. C. Batsel, Sept. 9, 1935, AEA, [25 376].

58. M. C. Batsel to Einstein, Sept. 10, 1935, AEA, [25 378].

59. Einstein to M. C. Batsel, [Sept. 13, 1935], draft, AEA, [25 380].

60. Einstein to M. C. Batsel, Sept. 13, 1935, AEA, [25 379].

61. M. C. Batsel to Einstein, Sept. 19, 1935, AEA, [25 381].

62. Einstein to M. C. Batsel, Sept. 22, 1935, AEA, [25 383].

63. Elsworth D. Cook to M. C. Batsel, Sept. 27, 1935, AEA, [25 385].

64. M. C. Batsel to Einstein, Sept. 27, 1935, AEA, [25 384].

65. Elsa Einstein to David Sarnoff, Oct. 29, 1935, AEA, [71 804], [51 591]. German and English versions of the letter are available at AEA. Because the English version has a date and locality, and its English is characteristic of Elsa Einstein's style, I quote this version, even though the German may be the original.

66. Einstein to David Sarnoff, May 20, 1936, AEA, [51 594].

67. David Sarnoff to Einstein, May 26, 1936, AEA, [51 594].

68. Einstein to Nathan Rosen, June 7, 1936, AEA, [89 307].

69. Einstein to Vyacheslav Molotov, Mar. 23, 1936, AEA, [20 212].

70. Engbert S. Reid to Helen Dukas, July 30, 1936, AEA, [51 595].

71. Einstein to Vyacheslav Molotov, July 4, 1935, AEA, [20 215].

72. US1992192: Gustav Bucky, "Oil Tank Level Indicator," 1935.

73. Gustav Bucky to Einstein, June 7, 1935, AEA, [35 427].

74. Einstein to Gustav Bucky, after June 7, 1935, AEA, [35 431].

75. Einstein to Gustav Bucky, [July 9, 1935], AEA, [35 432].

76. Einstein to Gustav Bucky, July 20, 1935, AEA, [35 434].

77. "Level Indicator," no date, AEA, [35 437].

78. Einstein to Gustav Bucky, no date, AEA, [35 435].

79. Einstein to Gustav Bucky, Nov. 3, 1936, AEA, [37 436].

80. Einstein to Gustav Bucky, July 20, 1935, AEA, [35 434].

81. US2058562: Gustav Bucky and Albert Einstein, "Light Intensity Self-Adjusting Camera," 1936.

82. Einstein to Gustav Bucky, [between Dec. 11, 1935, and Oct. 27, 1936], AEA, [35 477].

83. Einstein to Gustav Bucky, Nov. 3, 1936, AEA, [37 436].

84. New York Times, Nov. 27, 1936.

85. Sommerich to Gustav Bucky, Nov. 6, 1936, AEA, [35 475]. Patent Application in Germany for US2058562: Roefinag-Einstein-Bucky, [Dec. 1936], AEA, [35 476].

86. Einstein to Gustav Bucky, Mar. 9, 1936, AEA, [37 431].

87. Gustav Bucky to Einstein, Sept. 8, 1942, AEA, [35 483].

88. Einstein to Gustav Bucky, Oct. 8, 1942, AEA, [35 485].

89. GB523974: Leo N. Schwien, "True Air Speed Indicator," 1940. See also "Caltech Graduate Perfects True Air-Speed Indicator," Los Angeles Times, Oct. 27, 1940, B11.

90. Einstein to Gustav Bucky, [before Sept. 19, 1942], AEA, [35 430].

91. Gustav Bucky to Einstein, Sept. 19, 1942, AEA, [35 481].

92. Einstein to Gustav Bucky, [Sept. 21, 1942] Monday, AEA, [35 482].

93. Frank Aydelotte to Vannevar Bush, Dec. 19, 1941, National Archives, College Park, MD, Record Group 227, Records of the Office of Scientific Research and Development, UD Entry 11, S-1 Files, Bush-Conant Files, box 6, folder 72, Einstein, Albert.

94. Vannevar Bush to Harold C. Urey, Dec. 22, 1941, ibid.

95. Frank Aydelotte to Vannevar Bush, Dec. 24, 1941, ibid.

96. Harold C. Urey to Vannevar Bush, Dec. 29, 1941, ibid.

97. Vannevar Bush to Frank Aydelotte, Dec. 30, 1941, ibid.

98. For the FBI's investigation into Einstein's past and its denial of his security clearence to work in the Manhattan Project, see Fred Jerome, *The Einstein File: J. Edgar Hoover's Secret War against the World's Most Famous Scientist* (New York: St. Martin's Press, 2002).

99. Harold C. Urey to Vannevar Bush, Jan. 8, 1942, ibid.

100. Vannevar Bush to Harold C. Urey, Jan. 12, 1942, ibid.

101. Vannevar Bush, "Control of Gaseous Content," *Journal of the American Institute of Electrical Engineers* 41 (1922): 627–35.

102. Stephen Brunauer to Einstein, May 13, 1943, AEA, [81 005].

103. Frank Aydelotte to Stephen Brunauer, May 17, 1943, AEA, [81 007].

104. Einstein to Stephen Brunauer, May 17, 1943, AEA, [81 006)].

105. Stephen Brunauer, Paul H. Emmett, and Edward Teller, "Adsorption of Gases in Multimolecular Layers," *Journal of the American Chemical Society* 60 (1938): 309–19.

106. Burtron H. Davis and János Halász, "B.E. & T.: Scientists in the Background of Surface Science," *ChemTech* 21 (1991): 18–25; Burtron Davis, "Brunauer, Emmett and Teller–The Personalities Behind the BET Method," *Energeia* 5, no. 6 (1994): 1, 4–5; 6, no. 1 (1995): 1, 3–4.

107. Stephen Brunauer, "Einstein in the U. S. Navy," in Burtron H. Davis, and William P. Hettinger Jr., *Heterogeneous Catalysis: Selected American Histories* (Washington, DC: American Chemical Society, 1983), pp. 217–26.

108. Esther Caukin-Brunauer to Einstein, Apr. 22, 1938, AEA, [52 603]; Einstein to Esther Caukin-Brunauer, Dec. 18, 1938, AEA, [52 622].

109. Stephen Brunauer to Einstein, May 13, 1943, AEA, [81 005].

110. Stephen Brunauer, "Einstein and the Navy . . . 'an Unbeatable Combination,'" *On the Surface* 9 (Jan. 24, 1986): 1–2.

111. Ibid.

112. Irvin Stewart, *Organizing Scientific Research for War: The Administrative History of the Office of Scientific Research and Development* (Boston: Little, Brown, 1948), p. 85.

113. Stephen Brunauer to Einstein, May 21, 1943, AEA, [81 008].

114. Stephen Brunauer to Einstein, June 11, 1943, AEA, [81 009].

115. Frederick J. Milford, "US Navy Torpedoes. Part Two: The Great Torpedo Scandal, 1941–43," *Submarine Review* (Oct. 1996), www.geocities.com/Pentagon/1592 /ustorp2.htm?20075.

116. Einstein to Stephen Brunauer, June 18, 1943, AEA, [81 025].

117. William H. P. Blandy to Einstein, June 22, 1943, AEA, [81 011].

118. Lillian Hoddeson and Vicki Daitch, *True Genius: The Life and Science of John Bardeen* (Washington, D.C.: Joseph Henry Press, 2002).

119. Conference between Naval Ordnance Laboratory Representatives and Dr. Albert Einstein, July 2, 1943, AEA, [81 024].

120. Stephen Brunauer to Einstein, Aug. 12, 1943, AEA, [81 016].

121. Stephen Brunauer to Einstein, July 14, 1943, AEA, [81 013].

122. Einstein to Stephen Brunauer, July 16, 1943, AEA, [81 014].

123. Stanislaw Ulam, "John von Neumann, 1903–1957," *Bulletin of the American Mathematical Society* 64 (1958): 1–49.

124. Einstein to Stephen Brunauer, July 30, 1943, AEA, [81 027].

125. Stephen Brunauer to Einstein, Aug. 19, 1943, AEA, [81 017].

126. Einstein to Stephen Brunauer, Aug. 22, 1943, AEA, [81 032].

127. Roy W. Goranson to Einstein, Aug. 20, 1943, AEA, [81 018].

128. Frederic D. Schwarz, "Einstein's Ordnance," *AmericanHeritage.com History's Homepage. Invention & Technology Magazine* 13, no. 4 (1998).

129. Stephen Brunauer to Einstein, Aug. 28, 1943, AEA, [81 020].

130. Einstein to Stephen Brunauer, Sept. 1, 1943, AEA, [81 019], [81 029].

131. Einstein to Stephen Brunauer, Aug. 13, 1943, AEA, [81 028].

132. Stephen Brunauer to Einstein, Aug. 19, 1943, AEA, [81 017].

133. Einstein to Finkelstein, Oct. 27, 1943 (typed copy of the original), AEA, [78 939].

134. Schwarz, "Einstein's Ordnance."

135. Einstein to Stephen Brunauer, Jan. 1, 1944, AEA, [81 022].

136. Einstein to Stephen Brunauer, Jan. 4, 1944, AEA, [81 031].

137. Einstein to Stephen Brunauer and George Gamow, Oct. 15, 1944, AEA, [81 030].

138. George Gamow, *My World Line: An Informal Autobiography* (New York: Viking, 1970), pp. 149–50.

139. Brunauer, "Einstein and the Navy . . . 'an Unbeatable Combination.'"

140. Stephen Brunauer to Einstein, Aug. 19, 1943, AEA, [81 017].

141. Einstein to Stephen Brunauer, Aug. 22, 1943, AEA, [81 032].

142. Stephen Brunauer to Einstein, Aug. 12, 1943, AEA, [81 016].

143. Einstein to Stephen Brunauer, Aug. 13, 1943, AEA, [81 028].

144. Brunauer, "Einstein and the Navy . . . 'an Unbeatable Combination.'"

145. Ibid.

Index